Leading Together

Foundations of Collaborative Leadership

Curriculum for the Classroom, Grades 8-12

Laurie S. Frank
Carol Carlin
Jack Christ, Ph.D

with funding from
Sophia Foundation, Inc.

A Project of the Wisconsin Leadership Institute and the Collaborative Leadership Network

Published by:

Wood 'N' Barnes Publishing
2309 N. Willow, Bethany, OK 73008
(405) 942-6812

1st Edition © 2008, Wisconsin Leadership Institute

Cover Art by Blue Designs
Copyediting & Layout Design by Ramona Cunningham
Photographs by Belinda Ranstrom
Special thanks to the Quincy High School 2008 Business Technology Students of Quincy, MA who are pictured on the cover and interior.

Printed in the United States of America
Oklahoma City, Oklahoma
ISBN # 978-1-885473-79-0

The authors acknowledge permission to use the following articles and activities. We are especially grateful to those who allowed us to use their work, and who are also not authors of this curriculum. Access to their work makes this book richer, and will permit students and teachers access to a more well-rounded resource. We are grateful for their generosity.

Lesson 1.A
"Cards" description © 2006 Laurie Frank
"Group Bingo" description ©2003 Laurie Frank
"Name High Five" description ©2003 Laurie Frank
"Paired Activities" description ©2003 Laurie Frank
"Night at the Improv" description ©2003 Laurie Frank
"Cross the Line" description ©2004 Laurie Frank
"Handshakes" description ©2005 Laurie Frank
"Yurt Circle" description ©2003 Laurie Frank. Used with permission.
60 Second Speeches description ©2003 Laurie Frank
"Trust Walk" description ©2003 Laurie Frank
INTO THIN AIR by Jon Krakauer, copyright © by Jon Krakauer. Used by permission of Villard Books, a division of Random House, Inc.

Lesson 1.B
"Hand Slap" description ©2007 Marilyn Levin
Multiple Intelligences Indicator ©1999 Dell Coats-Erwin
"Categories" description ©2007 Laurie Frank
"Cliques" description ©2007 Marilyn Levin
"Find Your People" description ©2007 Laurie Frank
Chappelle, S. & Bigman, L (1998). Diversity in Action: Using adventure activities to explore issues of diversity with middle school and high school age youth. Beverly, MA: Project Adventure, Inc.
"Perspective" description © 2007 Marilyn Levin.
"Mind Power" description ©2007 Marilyn Levin
"Pairs Tag" description ©2007 Marilyn Levin
"Exploring Our Diversity Ups and Downs" description ©2007 Marilyn Levin
"Identity Line Ups description" ©2007 Marilyn Levin

Lesson 1.C
"Human Treasure Hunt" description © 2003 Laurie S. Frank
"Puzzles" description © 2003 Laurie Frank
"ABC Team" description ©2007 Laurie Frank
"Channels" description ©2003 Laurie Frank

Lesson 1.D
"Group Juggle with a Twist" description ©2007 Laurie S. Frank
"A What?" description ©2003 Laurie Frank
"1-1 Interview" description ©2003 Laurie Frank
"Interactive Video" description ©2003 Laurie S. Frank
Body Language Cues adapted from Givens, D. (2006) The Nonverbal Dictionary of Gestures, Signs & Body Language Cues. [electronic version] Center for Nonverbal Studies. Retrieved April 18, 2007 from http://members.aol.com/nonverbal2/diction1.htm
"All Toss" description ©2003 Laurie S. Frank
"Don't Spill the Beans" description ©2003 Laurie S. Frank
"Scamper" description ©2003 Laurie S. Frank
"Creativity: Four Measures of Style" description ©2007 Laurie S. Frank Adapted from Badger High School Leadership Curriculum (n.d), Lake Geneva, WI.
"Your Personal Network" description ©2007 Laurie S. Frank adapted from Badger High School Leadership Curriculum (n.d.), Lake Geneva, WI.
"Pathfinder Consent" ©2005 Laurie Frank
"Batten Down the Hatches" ©2003 Laurie Frank
"STEPS Decision Making process adapted from Badger High School Leadership Curriculum (n.d.). Lake Geneva, WI.
"Group Role Play" description © 2003 Laurie S. Frank
"Conflict Web" description © 2003 Laurie S. Frank
"Conflicting Needs" description © 2003 Laurie S. Frank
"Thumb Wrestling" description © 2003 Laurie S. Frank
"Moonball" description © 2003 Laurie S. Frank
"Warp Speed" description ©2003 Laurie S. Frank
"Nominal Group Process" description © 2007 Boris Frank. Boris Frank Associates. Paoli, WI.
"River of Life" description © 2003 Laurie Frank

Lesson 2.A
"Circle Scramble" description © 2003 Laurie S. Frank
"Toss 10" description © 2003 Laurie S. Frank
"Willow in the Wind" description © 2003 Laurie S. Frank
"Unification" description © 2003 Laurie S. Frank
"Activity Jumble" description © 2003 Laurie S. Frank
"Are You More Like?" description © 2003 Laurie S. Frank
"Commonalities" description © 2003 Laurie S. Frank
"People to People" description © 2003 Laurie S. Frank
"Screaming Toes" description © 2003 Laurie S. Frank
"Search and Rescue" description © 2003 Laurie S. Frank
"All Aboard" description © 2003 Laurie S. Frank
"Don't Touch Me" description © 2003 Laurie S. Frank

Lesson 2.B
"Think-Pair-Share" description © 2003 Laurie S. Frank

Lesson 2.C
"Continuum" description © 2004 Laurie S. Frank
Great Leaders Are Born; Great Managers Are Made by Ron Morris © 2003. Editor's Aide, Inc. Retrieved February 8, 2005 from http://www.imakenews.com/techyvent/e_article000118147.cfm

Lesson 2.D
"Leadership Egg Drop" description © 2003 Laurie S. Frank

Lesson 2.E
"Growth Circles" Description ©2003 Laurie S. Frank
"Lines of Vision" Description ©2006 Laurie S. Frank

Lesson 3.B
"Defining WLI Core Values" description © 2007 Laurie S. Frank
Lesson 3.C
"Balloon Frantic" description ©2003 Laurie S. Frank
"Classroom Parts" description © 2003 Laurie S. Frank

Self Assessment in Collaborative Leadership by Ruth A. Gudinas, FULL CIRCLE, N9136 Big Lake Road, Gresham, WI 54128-8955, 715/787-4427, with help from Laurie Frank.

"None of us is as strong as all of us."
—unknown

Contents

Foreword

It was exactly ten years ago that Laurie and I worked together on that first book about collaborative leadership, the *Manito-wish Leader's Manual*. What was true then is still true today. Students and teachers see leaders making decisions in their schools, communities, and the nation born out of expediency and made while managing endless distractions. But something is different today. I see men and women stepping up with fresh approaches to meet the challenges they face. Collaborative Leadership is catching on! If you want to join others in leading differently—if you wish to learn the fundamentals for getting something done—then this book is for you.

One of the best teachers I have had the privilege of being taught by was Jim Collins. In his book *Good to Great*, Jim writes about his observation of the greatest corporate leaders:

> *Level 5 leaders channel their ego needs from themselves and into a larger goal of building a great company. It's not that Level 5 leaders have no ego or self-interest. Indeed, they are incredibly ambitious—but their ambition is first and foremost for the institution, not themselves.*

Call it whatever you like, but Collaborative Leadership brings out the best in us and in others. It encourages giving without expecting repayment. Being a Collaborative Leader versus any other sort of leader is like the difference between being a tourist and a pilgrim. Tourists see and take, but pilgrims bring themselves and spend it on others.

I'm really not sure how we got to this point of needing Collaborative Leadership rather than any other kind of leadership. Perhaps "necessity is the mother of invention." Perhaps our decentralized, flat world demands that we work together to solve complex problems. Its use is counter intuitive because it can be messy at first, but my friend Fr. Bill Brennan S. J. reminds us that, "the greatest obstacle to good works is perfection."

So, does Collaborative Leadership work? The answer is an unqualified "yes." It does work and, more importantly, it must work. I know in my world, *partnerships* define the day. Our social networks often look like the butcher, the baker, and the candlestick maker—where everyone has their narrow area of expertise. How do we make decisions and meet challenges when we are so different and so competent as individuals? And, what will we do when we disagree? Collaborative Leadership holds the keys to bringing individuals together in a compatible social network in order to navigate challenges together. It is a powerful way to simply get things done in a world of increasing complexity.

There are other examples of *partnerships* in which Collaborative Leadership has been the context for success. In the social sector, donors are joining each other as partners to meet major challenges. Warren Buffett and the Gates Foundation now join together as colossal leaders in philanthropy.

Businesses see *partnerships* as essential to success. Central Florida attractions that compete for a family's vacation dollars decided that partnering to promote vacation packages including several attractions helped to build all of the businesses. After all, the real competition for all of them was the mouse!

Former rivals, Presidents Clinton and Bush, joined together to rally contributions for the Tsunami in 2005 and Hurricane Katrina relief in 2006. The results were stunning, drawing on the compassion of people from around the world.

If you are an action-oriented person and say to yourself, "I just want to get something done," relax, invest in teaching the skills found in this book, and then fasten your seat belts for accelerated action. Those who understand and use Collaborative Leadership are getting things done.

Writing a book can be a grueling task when it is not your "day job." Great leaders like Laurie, Jack, and the Wisconsin Leadership Institute Board didn't give up easily when distractions were everywhere. Imagine the thousands of unrelated new ideas that crossed their desks in e mails, telephone calls and meetings over two years of writing. Laurie and her team are a shining example of "stick-to-itiveness," as my mom used to say.

Bravo! You not only stuck to it but you have nailed the message and method for leadership.

May we plant a Collaborative Leader in every living room, classroom, and board room so that large challenges are made small and small challenges are made insignificant.

John L. Stanley
The Legacy Group, Inc.

Preface

Collaborative Leadership: How Do You Feel About People?

Over the long haul, truly effective leadership always starts and ends with a feeling for and about the people involved. Leadership always involves a group, team, organization, or community—in other words, other people. Leaders never accomplish anything alone. Thus it stands to reason that leaders should pay close attention to the people they influence and the people who might be helped or hurt by their decisions and their actions.

Unfortunately, leaders can be distracted by two factors that characterize every leadership situation: the goals they seek and the power they wield. And every time the group, team, organization, or community gets too big for people to work directly face to face, leaders may have trouble resisting the lure of large-scale goals or extraordinary power. Indeed, while extraordinary personal power is one of the most attractive enticements for individual leaders, it is also one of the greatest dangers for everyone else.

Leadership is about achieving goals, and sometimes people just have to get the job done regardless of how they feel about each other. On a battlefield or a sinking ship, for example, the way we feel about each other doesn't matter nearly as much as our immediate need to survive. And, because designated leaders have more power than other group members, it is often easy for them to neglect their responsibilities to other people and simply use their power for personal ends. Within families and other small groups where we mostly interact face to face, it's obvious that our behavior often depends on the way we feel about the other people. But ever since human beings created large groups in cities, then nations, and now global enterprises, the power of top-level leaders has grown way beyond natural proportions. When designated leaders lose direct contact with followers and others who are touched by their decisions, they are easily corrupted by their own power, especially if they have received that power as a birthright without actually earning it.

For thousands of years after people learned how to grow food, tame animals, and build large social structures, power and wealth were organized and passed on largely in family groupings, bolstered by legal systems, political and economic institutions, and armed force. Where these family groups retained a healthy fellow feeling for others in their community, their leadership tended to be effective and functional, and sometimes their communities thrived. Where these family groups got carried away by the lure of spectacular goals or unbounded power, their leadership sooner or later became dysfunctional and destructive, and their communities suffered.

The democratic revolutions of the last three centuries overturned or at least limited the dominance of powerful families and stressed the value of leadership by the most able, with the consent of the governed. In democratic cultures, the most important source of power became the power of widespread reliable knowledge rather than the power of brute force or manipulation of ignorance. If we are all to govern ourselves in a democratic community, then we must learn to trust each other as we generally do in small face-to-face groups and we must make sure that we all learn as much as possible in order to make wise decisions together. In fact, it was not until the era of the democratic revolutions that the word "leadership" emerged in our language. In previous eras, people spoke of rulership, kingship, or dominance, but the word "leadership" did not appear in our language until we learned to think about the widespread participation of the governed.

After five or six generations of democratic practice, sometimes at the cost of considerable struggle, three developments had bolstered a new optimism about people in general. First, our ideas about democracy had expanded to include people who were not full-fledged citizens in the original U.S. Constitution—African Americans and women. Second, research in the behavioral and social sciences had given us a much more positive view of human nature than we had previously recognized. And third, that research instigated a global movement toward self-improvement in virtually all aspects of the human experience. The leadership-development movement in general and the collaborative-leadership movement in particular may be the best examples. In important ways, the collaborative-leadership movement is a direct extension of the democratic spirit and the human capacity to learn, both individually and collectively. It recognizes the need for power and the importance of goal achievement, but it begins and ends with a simple and direct feeling for people.

Ironically, the roots of the collaborative-leadership movement are in the ancient world, especially in the wisdom of the great moral teachers who urged us to treat others as we wish to be treated and to love our neighbors as we love ourselves. Unhappily, those who have actually tried to practice what those prophets preached have all too often been cast as misguided idealists at best or as dangerous radicals at worst. From Socrates and Jesus to Abraham Lincoln, Mohandas Gandhi, and Martin Luther King, Jr., those who have actively preached compassion for and general faith in humanity have been violently scorned by many in their own time. It was not until the twentieth century, a century of spectacular violence and destruction, that the practical insights of popular opinion and scientific inquiry began to echo the truth of that ancient wisdom.

Since the 1950s, the work of many scholars in several disciplines has revealed that average, workaday human beings, under favorable conditions, are generally blessed with intelligence, willingness to work hard for relevant goals, an ample sense of responsibility, and potential for creativity. Until the twentieth century, only a small minority of human beings lived long enough and enjoyed sufficient leisure time to learn enough to demonstrate such potential. But by the middle of the twentieth century in the developed world, lots of people with advanced educations were living into their seventies and eighties and demonstrating that human potential is much greater than we ever realized before. Two students of human behavior during the second half of the twentieth century, Douglas McGregor and Robert K. Greenleaf, clearly demonstrate this development. Both helped to create a firm philosophic foundation for the principles of collaborative leadership.

McGregor was an industrial psychologist and a college president who observed thousands of executives and managers in action. In *The Human Side of Enterprise*, published in 1966, McGregor concluded that all of the executives and managers he studied generally subscribed to one of two distinct attitudes about normal average human beings. He called these two attitudes Theory X and Theory Y. Theory X managers tended to believe that most people most of the time are self-centered, childish, lazy, not very intelligent, not creative, and not very willing to accept responsibility of any sort. Theory X managers believed that only a small fraction of the human population has what it takes to be leaders, and that this small fraction was born with the inherent traits needed to be leaders. In other words, they believed that leadership skills are not learned but are always born into those we call leaders. Theory X managers have very little faith in people other than their own circle of family members and associates. Their management strategies are not really about leadership at all, but rather about the need to control, coerce, and manipulate people. To a large extent, Theory X managers resisted the central principles of democratic self-governance based on faith in average people. McGregor's name for Theory X management was "management by control."

If we look over the whole sweep of human history, we can identify Theory X attitudes in lots of times and places, especially before the democratic spirit arose. All slave systems were essentially Theory X in that they failed to see anything in run-of-the-mill humanity other than a brute pack animal. One key to the Theory X mentality is that it interprets most other people as a means to one's own ends or as a tool to achieve one's own goals rather than as ends in themselves. Thus goals and power take priority over concern for people, and workers become interchangeable or disposable. To be fair, it's easy enough to find evidence supporting Theory X over most of human history. Where most of the world's work is physical, dirty, and disagreeable, and where virtually everyone is thoroughly ignorant, people who were forced to do all that dirty work probably behaved like pack animals or self-centered, lazy, irresponsible children—at least in the eyes of those who made them do that work. Even today, the dirtiest jobs are generally performed by the people with the least education and the least opportunity for better work, and it is hard to manage or motivate people in such situations without some measure of coercion, manipulation, and control.

McGregor also described a growing number of managers who practiced what he called Theory Y management or "management by objectives." These managers tended to see the upside of human nature, the potential for creative work and responsible decision-making. Such managers believe that average people are reasonably intelligent, potentially creative, willing to work for goals which relate to their own welfare, and eager to accept responsibility for things they feel are important. Theory Y, however, is not just a reverse image of Theory X. Where Theory X managers see a lack of intelligence, Theory Y managers recognize that intelligence comes in various forms and that different people are smart in different ways. They also recognize that creativity comes in different packages. As for a work ethic and a sense of responsibility, Theory Y managers recognize that people should not be expected to work hard or to take responsibility purely for somebody else's benefit. According to a Theory Y perspective, people will naturally seek responsibility for their own welfare and the welfare of people they care about. Theory Y managers don't threaten punishment or rely strictly on promises of reward; instead, they draw out the intelligence, creativity, and commitment of group members by helping them clarify their shared objectives and by supporting them in their work to achieve those objectives. In other words, Theory Y managers and leaders help people collaborate. Historically, the move from Theory X to Theory Y is a move from traditional top-down authoritarian leadership to collaborative leadership.

We can find hints of Theory Y sprinkled throughout the human record, though, as we've seen, Theory Y has traditionally been associated with radical idealism and unrealistic wimpiness. The message of the world's major religions to love each other and recognize our universal kinship is a Theory Y idea. Political and social democracy is a Theory Y idea: if we really think that people should govern themselves, then we must trust people to be intelligent, responsible, and even creative. Thus the best thinking has always been Theory Y, but we haven't been able to realize that until recently, after lots of people reached a stage of relative prosperity, adequate leisure, and advanced education. The new industries which depend very heavily on education and creativity (e.g., the computer industry and the education industry itself) require the encouragement of individual responsibility, creativity, and teamwork. For a couple of generations now, economists and social scientists have agreed that the most important economic resource is no longer land or money, but widely distributed knowledge that can be put to use in networks of collaboration.

When all is said and done, Theory X and Theory Y are both "self-fulfilling prophecies." In other words, they both become true precisely because one believes they are true. For example, the stock market rises when people *believe* it will rise (thus motivating them to buy stock, which actually causes stock prices to rise). And the stock market falls when people *believe* it will fall (thus motivating them to sell stock, which automatically depresses the price). Similarly, people behave in Theory X or Theory Y ways depending on how we expect them to behave (and thus how we treat them). The classic research on this came about through an accident in a British school system. A teacher was mistakenly told that her elementary students were backward and came from homes where education was not valued. The teacher tried to be a good teacher, but she unwittingly treated the students as if this description were true. On subsequent test scores, the children did poorly. Then somebody realized the mistake and told the teacher that her students were actually quite bright and that their families supported their learning. The teacher started treating the students as if they were bright and eager, and, OOPS, they all brightened up and got eager. On the next round of tests, they scored quite well. Good parents, good teachers, and good managers know that people respond in kind to the way they are treated.

The "servant leadership" movement instigated by Robert K. Greenleaf illustrates some of these same themes. When Greenleaf graduated from college in 1925, he dreamed of making an important contribution to the world, and he believed that his best chance to do that was in a large corporation. So he went to work for the American Telephone & Telegraph Company, then the world's largest and probably most effective business. He worked his way up from the bottom to the position of Vice President and Director of Management Research, a role he created for himself that gave him lots of room to explore his own ideas about organizational management and leadership. AT&T gave him extensive freedom to study leadership and management behavior in an attempt to develop the best possible leaders for the company.

After years of research and observation, Greenleaf concluded that the best leaders start with a desire to serve rather than a desire for power or goal achievement. He also proposed that this desire to serve begins with a simple and natural feeling for people. Near the end of his life, he published a short pamphlet, *The Servant as Leader*, in which he stated the following:

> The servant leader is servant first. It begins with the natural feeling that one wants to serve, to serve first. Then conscious choice brings one to aspire to lead.... The difference manifests itself in the care taken by the servant—first to make sure that other people's highest priority needs are being served. The best test, and difficult to administer, is: do those served grow as persons; do they, while being served, become healthier, wiser, freer, more autonomous, more likely themselves to become servants? And what is the effect on the least privileged in society; will they benefit, or at least not be further deprived? (1991, p. 7)[1]

The spirit of Greenleaf's message has grown significantly, thanks largely to the work of the Greenleaf Center for Servant Leadership in Indianapolis, though the message has not been widely noticed in mass media. As usual, the top-down authoritarian model of leadership still seems more

[1] Greenleaf R. K. (1991). *The Servant as Leader.* Westfield, IN: The Robert K. Greenleaf Center.

practical and more attractive to power holders and those who long for power. Greenleaf himself anticipated this problem and believed that getting the natural servants to step up and take leadership roles would be even more difficult than getting the power holders to let go of their power. In this dilemma lies the paradox of collaborative leadership. Those with a natural desire to serve must learn to lead, while those with a desire for power must learn to serve. Traditional leaders must learn to step back and servant leaders must learn to step up.

We need collaborative leadership now more than ever. As the world has shrunk to the size of a laptop computer, and as lethal weapons have become more powerful and more widely available, the primary challenges for leaders in the twenty-first century will be to tear down walls and build bridges; to listen carefully; to work, play, and learn with each other; to model peaceful, patient means to achieve communal ends; to minimize the influence of fear and anger as motives for action; to promote stewardship of our natural resources; and to create communities of reciprocal respect, care, and responsibility. In other words, leaders in the twenty-first century must learn to collaborate.

Luckily, our need for collaborative leadership has emerged while enormous new opportunities for collaboration have appeared, almost overnight. Stunning advances in scientific knowledge and technological development like the Human Genome Project, the open-source movement, and the global networks available on the Internet have all drawn on the communal capacity for invention, innovation, and collaboration. Old forms of top-down authoritarian leadership and self-centered notions of property ownership are giving way to free-wheeling, co-creative collaborations in business, education, nonprofit enterprise, and even government. Where such need and such opportunity meet, wonderful things can happen. And if we can help young people learn the skills and habits of collaborative leadership, they can make wonderful things happen for the benefit of all humanity down through the generations.

Jack Christ
Executive Director of the Wisconsin Leadership Institute
Professor of Leadership Studies at Ripon College

Acknowledgments

This curriculum on collaborative leadership required years of collaboration from a multitude of talented and committed volunteers. The people who helped bring this topic to life are passionate about changing the paradigm of leadership in the world. They are equally passionate about helping young people develop into servant leaders who believe in their own capacities and have faith in their fellow humans to co-create a world filled with compassion and integrity. Because collaboration invites collaboration, the ripple effect continues through this curriculum. The authors are indebted to those who have contributed. By throwing their personal stone in the waters, they helped to make the ripples travel farther and faster.

The central principles of this curriculum were identified by the Collaborative Leadership Network (CLN) curriculum development team in 2001. We began with the assumption that any attempt to develop effective collaborative leadership must be inspired by a clear sense of the roles and responsibilities of an effective collaborative leader in all times and places. The list below is the result of that work and the foundation of the curriculum itself.

The work was facilitated by Judy Neill of the Worldwide Curriculum Design System, aided by her colleague Dick Zellmer. The CLN team included Laurie Frank, CEO of Goal Consulting and primary author of this resource; Dr. Jack Christ, Executive Director of the Wisconsin Leadership Institute and Professor of Leadership Studies at Ripon College; Terry Shelton, Director of Outreach for the LaFollette School of Public Policy at the University of Wisconsin-Madison; Dr. Douglas Northrop, Executive Director of the Ethical Leadership Program at Ripon College; and Jan Burt, Program Director for the Wisconsin Center for Academically Talented Youth. The team spent most of 2001, including two full-day, face-to-face working sessions, brainstorming, organizing, and honing the list of roles and responsibilities essential to effective collaborative leadership.

These principles guided the drafting of the curriculum itself. Laurie Frank did most of that work with collaboration from Jack Christ and Carol Carlin, who taught a very successful leadership course at Badger High School in Lake Geneva, Wisconsin, for ten years. Thanks to a series of grants from the Sophia Foundation of Fond du Lac, Wisconsin, the introductory segments of this curriculum were completed and introduced to secondary-school teachers and students at two two-day workshops held at Ripon College in the summer of 2004. Based on their workshop experiences, several teachers in Wisconsin and Illinois agreed to test the curriculum in their own schools during the 2004-2005 school year. That testing and piloting period proved successful and contributed to the revision of the curriculum that appears in this first edition. Those who helped in this effort were:

Diana Aguilar	Student, Morton High School, Cicero, IL
Juan Aguilar	Student, Morton High School, Cicero, IL
Hellen Bermudez	Student, Morton High School, Cicero, IL
Sue Brereton	Assistant Principal, Ripon High School, Ripon, WI
Gene Delcourt	Teacher, Shabazz High School, Madison, WI
Roman Emano	Teacher, J. Sterling Morton High School, Cicero, IL
Carla Hacker	Stress/Challenge Coordinator, Madison Metropolitan School District, Madison, WI

Jody Henderson-Sykes	Teacher, Longfellow Schools, Milwaukee, WI
Mary Hunter	Green Lake High School, Green Lake, WI
Suzanne Katz	Ripon College, Ripon, WI
Andy Kirchmeier	Ripon College, Ripon, WI
Ali Nowak	Ripon College, Ripon, WI
Becca Nowak	Ripon College, Ripon, WI
Edgar Pacheco	Student, Morton High School, Cicero, IL
Rafael Ramirez	Morton High School, Cicero, IL
Jose Luis Torres	Student, Morton High School, Cicero, IL
Jeanne Williams	Ripon College, Ripon, WI

All of this brilliant collaboration would not have been possible without the vision, foresight, and commitment of the Wisconsin Leadership Institute board of directors. Volunteers all, they have, over the past half-dozen years, guided this project to fruition. These people have been the mid-wives of this process, modeling the patience and persistence necessary in genuine collaboration:

Zena Bauer, Student, Ripon College
Polly Beal, Fund-raising and Public Affairs Consultant
Katherine Burnham, Founder and President, Society for Nonprofit Organizations
Jan Burt, Program Director, Wisconsin Center for Academically Talented Youth
*Anne Derber, Executive Director, Camp Manito-wish YMCA
Virginia Duncan Gilmore, Business Leader, Community Volunteer, Philanthropist
Kelly Haverkampf, Executive Director, Wisconsin Rural Partners
Nancy Johnson, Vice President, Corporate Research, American Family Insurance
Garth Katner, Education Director, Interfaith Youth Core
*Melinda McNett, Student, Ripon College
David Minor, Director of Corporate and Government Relations, Ripon College
Megan Piotrowski, Student, Ripon College
Paul Ranslow, President, Ripon College
*David Seligman, Executive Director, Ethical Leadership Program, Ripon College
*Terry Shelton, Outreach Director, LaFollette School of Public Affairs, University of Wisconsin-Madison
David Simmons, Information Technology Consultant
John Stanley, Executive Director, Camp Manito-wish YMCA; Founder and President, The Legacy Group
*JoAnn Stormer, Executive Director, Wisconsin Rural Leadership Program
Rolf Wegenke, Executive Director, Wisconsin Association of Independent Colleges and Universities
Jennifer Welsh, Assistant to the Governor of Wisconsin
*Mark Zanoni, Director of the Leadership Program, Camp Manito-wish YMCA

June Jordan said that "we are the people we've been waiting for." This curriculum is designed to help young people realize their own leadership potential. It was created by so many for those who are ready to stop waiting and begin their leadership journey.

---

* Indicates current member. (Board Presidents in chronological order are Duncan Gilmore, Burnham, Shelton, Stormer.)

Introduction

When John Stanley called and asked if I might be interested in being a partner in developing his vision for a leadership program for youth at Camp Manito-wish, I wasn't sure if I should accept or run as fast and as far as possible in the other direction. As with many people, my baggage around the idea of leadership lay in the lessons learned as a child, best summed up on a common t-shirt message: "If you are not the lead dog, the view is always the same." I grew up believing that leaders are loud, charismatic people who find a parade and jump in front of it. Although I had been collaborating with people all my life, I had not entertained the notion that leadership could, in fact, be shared. Here I was in my late 30's and I was being told that leadership could be undertaken in groups, where people can use their strengths in a dance of give-and-take, and where everyone is seen as equal and contributing members. Now this I could get into. This idea of *collaborative* leadership was right up my alley.

After 15 years of collaboration with John and the folks on the board of the Wisconsin Leadership Institute, I have a new appreciation for the possibilities of leadership embedded in collaboration. Leadership, in this sense is about how the individual acts out leadership opportunities in the social context. It is about using the strength of relationships—the power of people working together—to make change, achieve goals, and simply get things done. The core of collaborative leadership lies in the following foundational beliefs:

- **Everybody has the capacity to lead.**
  Leadership can be learned. People can lead in all parts of their lives—families, business, community, church groups, athletic teams, camps or clubs—there are many opportunities for leadership. This idea runs counter to the stereotype of leadership where one must be special, elected, or appointed in order to be a leader.
- **Increasingly, leadership will come not from a position of authority, but from within a group.**
  As our world expands and becomes more complex, there is a trend toward decentralization. It is imperative that young people gain skills to operate in small groups and as members of teams.
- **Leadership takes place within the context of relationships.**
  Collaboration is more than working together. It involves pooling the strengths of all individuals in order to create something that could not be accomplished alone. In order to truly collaborate, it is essential to build trust between the individuals in the group.
- **In order to learn how to lead, one must have the opportunity to lead.**
  Learning to lead is an experiential process, requiring skills that take practice. Talking about leadership is not enough. People must have opportunities to practice leadership in a safe environment, where they can learn from their mistakes.
- **Leadership involves risk taking.**
  Choosing to step up, give an opinion, try new ideas, and listen to those with whom one disagrees are acts of risk taking. Therefore, young people need opportunities to examine their own beliefs and reactions around the taking of risks.

- **Leadership requires an action orientation.**
  There is a subtle difference between being a leader and not being a leader. It involves stepping forward—taking the initiative—looking for what needs to be done, and being willing to do it. Leadership also entails modeling.

- **Learning to be a leader is a lifelong journey that begins with the question, "What is leadership?"**
  Everyone defines what leadership means for him or herself, and this definition can change over time as one becomes more aware and experienced with the idea of collaborative leadership.

Bill McKibben, who wrote the book *Deep Economy*, spent a year watching every show he could find from cable television. In an interview on radio's "To the Best of Our Knowledge" he shared his learning from that experience:

> The central message of the consumer culture in which we live is: You're the most important thing on earth. You're the heaviest object in the universe and everything orbits around you. And we've enshrined this idea as "human nature." Not remembering that most people in most places have had other things very near the center of their identity—the tribe, the community, their relationship with the natural world, or the Divine—something that gave them more of a sense of identity not obsessively rooted in themselves.*

It may be that collaborative leadership runs counter to the dominant message in the United States. It may also be that learning to lead collaboratively can bring us, and our youth, back in touch with the "other," and expand each person's world outside of his or her immediate domain. And it may be that collaborative leadership, by harnessing the varied strengths of those within the various groups is one of the answers to bringing the world back from the brink of disaster from global warming, genocide, famine, and a host of other worldwide ills. Who knows? We certainly have nothing to lose and everything to gain. The best part is that our youth can grow up to be part of the solution.

Laurie Frank
August 2008

---

* Wisconsin Public Radio (interview aired May 26, 2007). *To the Best of Our Knowledge*. Public Radio International: Minneapolis, MN.

Using This Curriculum

If you choose to teach this curriculum, it is helpful to have an affinity for the philosophy implied by the foundational beliefs. In other words, if you concur with the stated beliefs, or bias, in this curriculum, it will make much more sense to both you and your students as you work through the process together. The purpose of this section is to make the inherent biases in this curriculum transparent by offering definitions, and then comparing our viewpoints on traditional versus collaborative leadership. We conclude with a perspective on the role of reflection in the learning process.

At this point in history, the purposes of leadership may be more important than any particular definition of leadership. Nonetheless, in an introduction to a book about collaborative leadership, some attention to definitions is useful and even necessary. Thus we propose the definitions noted below for leadership in general and for collaborative leadership in particular. Both are reciprocal processes; in other words, leaders and followers always act in response to each other and both influence each other in various ways. Leadership in general and collaborative leadership in particular both arise from the human need to solve new problems and create new tools or new solutions to problems. Thus they also require clarification of group goals, communication of strategies for goal achievement, organization of resources, and acceptance of responsibility for results. They differ in their comparative focus on goal achievement or relationships and in the scope of their vision.

> **Leadership** is a reciprocal process of encouraging and supporting people in the pursuit of goals shared by members of a group, organization, or community.
>
> **Collaborative leadership** is a reciprocal process of encouraging and supporting relationships within which people can pursue a variety of shared goals over extended periods of time.

In this context, the word "encouraging" must be interpreted in its broadest and deepest senses. "To encourage" ultimately means "to give courage." Collaborative leaders give courage mostly by setting examples, modeling behavior, and taking creative personal risks for the good of the group. Collaborative leaders don't necessarily ask others to follow them but to join with them. In doing so, they act out Gandhi's injunction to "be the change you want to see in the world."

In conjunction with the above definitions, it is helpful to compare the historical (or "traditional) views of leadership, and this relatively new style of leadership that we are calling "collaborative." The charts on the following pages identify a variety of characteristics by which collaborative leadership behaviors, styles, and tendencies differ from the traditional leadership models.

Comparisons Between Traditional and Collaborative Leadership Behaviors

The following charts identify a variety of characteristics by which collaborative leadership behaviors, styles, and tendencies differ from the traditional leadership models that held sway for centuries before the advent of democratic forms of social organization and throughout the eras of agricultural and industrial economies. Because it's a chart, it is often oversimplified and in need of explanation and clarification. Nonetheless, it can be useful as a general guide to understanding the differences between the old and new ways of doing leadership.

OPERATING ENVIRONMENT

Characteristic	Traditional Approaches	Collaborative Approach
Market forces	Stability	Rapid change in technology, society
Time frame	Short-term objectives	Long-term vision, mission, values
Focus	Bottom-line, product	Shared mission, win/win
Operating norms	One-time events, crisis	Ongoing relationships, processes
Problem analysis	Simple cause-effect	Overlapping boundaries, systemic
Work structure	Division of labor	Cross-disciplinary teams
Work process	Bureaucracy, rules, regulations	Networks, shared visions, long-term mission focus
Change	Driven by necessity, crisis	Driven by innovation, continuous learning
Impact	Local	Global

RELATIONSHIPS

Characteristic	Traditional Approaches	Collaborative Approach
Power structure	Hierarchies	Networks, communities
Power flow	Top-down	All directions
Power relationships	Command, control	Service to others
Authority	Received authority	Fluid authority, authenticity
Goals, ideals	Efficiency through routine, mechanization	Innovation through creativity, lifelong learning, renewal
Leadership focus	Recognition of position, people as means to ends	Recognition of individual, people as ends in themselves
Leadership approach	People as strictly physical creatures	People as complex spiritual beings
Leadership tactics	Fear, manipulation, charisma	Example, empowerment, persuasion, humor, wisdom
Impact	Local	Global

PROCESSES		
Characteristic	**Traditional Approaches**	**Collaborative Approach**
Work environment	Separation, segregation, self-centeredness	Integration, empathy, compassion for others
Participation	Homogeneity	Diversity
Education	Formal, separate from work life	Continuous, life-long, integrated with work life
Work values	Succeed or fail	Experiment, learn
Work relationships	Self-reliance, autonomy	Interdependence
Problem solving	Strictly linear, logical	Acceptance of paradox, ambiguity
Interaction with environment	Instrumental, exploitative	Based on stewardship, integrity
Management style	Risk reduction	Responsible risk
Reward structure	Immediate goals	Long-term comprehensive learning
Outcomes	Either/or, win/lose	Both/and, win/win

PERSONAL VALUES, BEHAVIORS		
Characteristic	**Traditional Approaches**	**Collaborative Approach**
Projected image	Invulnerability, physical courage	Vulnerability, flexibility, growth, intellectual/emotional courage
Orientation to people	Low regard for average person	High regard for universal human potential
Communication style	Talk, give orders, answer questions	Listen, consult, ask questions
Motivating tactics	Reward, threaten, demand compliance	Discern others' needs, coach, facilitate, generate commitment
Information exchange	Transmit data	Tell stories
Problem solving	Address crises, solve obvious problems	Discover, meet unstated needs
Management style	Risk reduction	Responsible risk
Reward structure	Immediate goals	Long-term comprehensive learning
Outcomes	Either/or, win/lose	Both/and, win/win

Thinking It Over: The Role of Reflection

Many students of leadership behavior have noted that the best way to develop leadership skills and values is to reflect on one's own experience. That is certainly true of this curriculum on collaborative leadership. The exercises, readings, and activities are intended not only to be completed, but also to be reflected upon. This raises an obvious question: what is the role of reflection in the learning process?

In a literal physical sense, "reflecting" implies giving back an image, as a mirror or a puddle reflects light from a source. The prefix "re" generally means "to repeat" or to do over again; so when we see our reflection in a mirror or a puddle, we are not seeing our actual self but seeing our image "over again." Reflecting also implies thinking, of course, so when we reflect or think about something we are sort of living it over again. When we "think something over" or "mull something over," we are reflecting again and again. In this sense, reflection is the act of thinking it over, or thinking again, or even thinking it over again.

The act of reflecting on one's own experience is actually something of a miracle, or at least a puzzle. How can anyone be both the object of reflection and the one doing the reflecting? This puzzle is at the heart of individual personal mastery, which is a big part of the challenge of leadership development. Leaders must first be able to lead themselves before they can lead others. And they learn to lead themselves largely by taking on leadership challenges and then reflecting upon them. By thinking things over and over, aspiring leaders actively make their experience their own rather than passively letting it happen to them. In doing so, they create meaning and purpose for their own futures.

Thinking your experience over and over without changing your viewpoint on it may help to solidify the experience in your memory, but that won't necessarily help you learn the larger lessons of the experience. You must also understand what kinds of different questions you can ask about the experience to make sure you aren't just reinforcing the wrong conclusions based on faulty perceptions, faulty information, or faulty interpretations. **In general, it's a good idea to think about any experience or any idea from five different perspectives: available facts, emotional responses, threats or dangers, optimistic dimensions, and creative potential**.

If, for example, you want to make sense of your experience of reading about the adventures of Ernest Shackleton and the crew of the *Endurance*, you can start simply by making note of all the **verifiable facts** of the story. What do we know about the men who attempted to conquer the South Pole, about their goals and purposes, about what it would take for anyone to succeed in such an effort? What do we know about the impact of their adventure on those who read about it afterward? Available facts can be based on people's opinions, by the way, as long as we know what those opinions really are.

Emotional responses simply refer to how you or others feel about the various aspects of the experience. What emotions did you feel as you read the story? What emotions do you imagine the participants were experiencing as their adventure was unfolding? What emotions do you think their story evoked in those who heard about it later? In one respect, the available facts and the emotional responses overlap: the verifiable fact that the story of Shackleton and the *Endurance* has been told to millions of people over several generations, indicates that emotional responses to the story have been substantial for quite some time.

Thinking about the **dangers and threats** implied by this particular story is pretty easy, especially because the human brain has evolved largely to identify threats and make quick decisions about how to respond. Our instinctive reactions to danger are to fight, flee, or freeze. In response to physical challenges like the story of the *Endurance*, these responses make perfect sense; but in our modern civilized world, most of the challenges we face call for more complex and sophisticated responses. In the natural world where physical danger lurks all around us, this dimension of thinking is vital. It is also useful to us now, but it should not dominate our reflections. The default mechanisms in our brains make us focus on the negative and pessimistic dimensions of experience, so we may have to work hard to focus on the positive and optimistic dimensions. Focusing on the positive and optimistic dimensions, by the way, is one of the most important duties of leaders, especially when times get tough. In any case, imagining ourselves in the shoes of Shackleton and his crew and reflecting on the dangers and threats they faced, can help us learn strategies and tactics to apply to analogous situations in our own experience.

The adventure of the *Endurance* crew may have been extraordinarily grim and at times even depressing, but reflecting upon it should not ignore the possibility of **positive, optimistic lessons**. We do, after all, wind up marveling at the courage, persistence, strength, and teamwork of the men and the fact that they all survived virtually impossible conditions. In this case you could also reflect on the strategies and tactics Shackleton himself used to keep his men from sinking into depression and despair. For that matter, you could reflect on the fact that Shackleton was able to keep his own spirit from sinking.

Finally, every story and every experience can serve to trigger connections to other stories and other experiences. Making connections is the central act of creativity, so reflecting on everything in the *Endurance* story to imagine how they might connect with other stories or with experiences from your own life can be very useful. Another way to think about this, is to ask yourself how you can apply the lessons of experience to other lessons from other experiences and to potential future experiences. The primary resource for this kind of **creative reflection** is imagination, mixed with memory, analysis, and aspiration.

Collaborative leadership is ultimately a creative act. The process of reflection that helps aspiring leaders to develop their own skills and values culminates in imaginative creativity. Along the way, it makes use of reflection about available facts, emotional responses, threats and dangers, and sunny optimistic interpretations. As you work your way through this curriculum on collaborative leadership, please take ample time to reflect on each step of the journey, to think your experiences over, then think about them again, then think them over again.

How to Use This Curriculum

There are many challenges to writing any curriculum, and this one proved no different. Aside from providing the necessary detail, and targeting of an audience (in this case, grades 8 to 12), it is important to recognize that every school/class/teacher/system is unique. As such, every person picking up this resource is the acknowledged expert in his or her situation. Each person will use her or his own discretion about where to start, which activities to experience, how to introduce concepts, and which tasks will be done in class or as outside assignments.

In addition to the standard challenges, this curriculum provided some special conundrums. A particular determination must be made about how much time to devote to each lesson. Because collaboration is based in relationships, it is necessary to provide time to develop them. When assigning time frames to the lessons, we had to ask ourselves, "how much time does it take to create constructive relationships?" The answer, of course is, "it depends." Some people connect easily, while others have to intentionally struggle at developing working relationships.

As a general time frame, Unit 1 can certainly be accomplished in a semester of 45-minute classes, or a quarter in a 90-minute block schedule. It is possible to combine Units 2 and 3 in one quarter, although it is suggested to separate them into two quarters, which allows more time to develop the concepts on a deeper level. Much of the time frame is determined by a variety of factors best left to the discretion of the teacher who knows her or his students, schedule, and situation. Some parts of any given lesson can be accomplished in class or assigned as homework. When choosing how much time to devote to a given activity, please remember that there are many opportunities for teachable/learnable moments. A discussion may take a surprising and stimulating turn, or an activity may provide more questions to mull over together than are presented in these pages. Likewise, something that appears time-consuming on paper may end up taking less time, or the focus is such that the teacher chooses to extend a journaling session.

In short, this whole curriculum can be completed in one year. It can also be divided into levels so that different grades experience different parts of the curriculum. For example, grades 8 and 9 may focus on Unit 1, while grades 10 through 12 delve into units 2 and 3. In an ideal world, time is not an issue, and you are able to move through the concepts at your leisure, taking as much time as necessary for each to be developed to its fullest.

The other conundrum involved the nature of experiential learning. Educators are well aware that learning is not a linear process. Please note that experiential learning is unpredictable, often enjoyable, and generally messy. The challenge, here, was to write a rather abstract process in a linear fashion, the result of which you hold in your hands.

This curriculum uses an experiential process as the methodology for teaching collaborative leadership. As such you will notice that each lesson contains at least three (and generally more) steps. Activities, in and of themselves are not experiential. If they are used as vehicles from which to reflect, dialogue, and apply concepts and skills, then the activities are part of an experiential process.

Simply doing the activities will probably not allow the individuals and the group to progress with collaborative leadership. Most of the learning will be the result of the group discussions, journal entries, and projects that grow out of the activities. Likewise, having discussions and reflection without the activities may create a situation where the concepts are too abstract to be meaningful for many people. Together, the steps of the lessons create rich and varied learning opportunities for students.

Ideally, you will have the opportunity to use the whole curriculum from beginning to end. The skills are scaffolded so that one lesson builds on the other. Also note that some of the lessons refer to earlier exercises in the curriculum. Each unit can also stand alone, depending on the needs and outcomes of the students with whom you work. If, for example, you wish to help your class become more collaborative, Unit 1: What is Collaboration?, is a valuable resource. If, on the other hand, your students have been together for a long time and are already using the collaborative skills outlined in Unit 1, then a natural starting point is Unit 2, "What is Leadership?" Finally, if you are working with a group of seniors who have taken on leadership roles, and worked collaboratively for awhile, they may be ready to dig into the essence of collaborative leadership in Unit 3, "What is Collaborative Leadership?"

Whichever scenario or starting point you choose, the authors suggest that you always begin with lesson 1.A: Develop Group Cohesion to make sure the group has a working social contract and students are prepared for their collaborative journey. It is also strongly suggested that you refrain from picking and choosing isolated activities, and use the full range of reflection, generalization, and application exercises that go along with the activities.

Below is an overview of the curriculum. The titles are posed as questions because learning about collaborative leadership through this curriculum is a constructivist process, whereby students construct meaning through the experiences you provide. Although the Wisconsin Leadership Institute and the authors of this curriculum advocate a particular philosophy around leadership development, and information from the broad field of leadership studies is presented, there are no wrong answers. Collaboration is steeped in dialogue, disagreement, listening, and open-mindedness. Your class can live the collaborative model.

Course Overview

Unit 1 Lessons	**What is Collaboration?** 1.A Develop Group Cohesion 1.B Recognize the Diversity Within the Group 1.C Differentiate Between Cooperation and Collaboration 1.D Use a Variety of Strategies to Maximize Collaboration Within a Group 1.D[1]: Communication 1.D[2]: Decision-Making 1.D[3]: Conflict Resolution 1.D[4]: Goal-Setting 1.E Apply Collaborative Skills
Unit 2 Lessons	**What is Leadership?** 2.A Create Community 2.B Compose a Personal Definition of Leadership 2.C Compare and Contrast the Notions That Leaders are Born and/or Made 2.D Explore Different Views of Leadership 2.E Analyze the Roles of Risk Taking and Long-Term Vision When Leading 2.F Construct a Vision of an Ideal Leader
Unit 3 Lessons	**What is Collaborative Leadership?** 3.A Continue to Intentionally Build and Maintain a Safe Working Environment 3.B Assess Personal Collaborative Leadership Values 3.C Use Collaborative Leadership Qualities and Values 3.D Practice Effective Use of Collaborative Leadership Core Values and Qualities

Roles and Responsibilities of Collaborative Leaders

Collaborative leadership is not easily described. There is a depth to it that is difficult to get one's arms around. People who are adept at leading collaboratively are generally admired for their proficiency in facilitation, listening, nurturing relationships, and creating shared visions. How can these complex skills be boiled down to parts that are concrete enough to teach to others?

The Collaborative Leadership Network did just that. In 2001, a group of people sat around a table at the Worldwide Instructional Design System (WIDS)* office in Middleton, Wisconsin discussing the nature of collaborative leadership. Under the deft facilitation of Judy Neill, they teased out the bits and pieces of duties, or tasks, a person needs in order to skillfully practice collaborative leadership.

At some point, all the roles and responsibilities noted below become central to effective collaborative leadership. Unfortunately, no single individual is likely to be especially gifted in all of them, and nobody can act them all out at once. Fortunately, part of the genius of collaborative leadership lies in its faith in the wisdom and judgment of the group as a whole. While no single person may be really good at all these roles and responsibilities, together group members can probably play all the roles necessary when the needs arise. The secret is to create an environment that encourages all to participate, to be mindful of one's role as a group member, and to remember that learning about collaborative leadership is a process of never-ending growth.

Many of these roles and responsibilities are directly addressed in this curriculum (***indicated in bold italics***) and create the focuses for each lesson. Those that are not dealt with here will be the subjects of future, more advanced, curricula.

ROLES AND RESPONSIBILITIES OF COLLABORATIVE LEADERS

Promote collaboration

- ***Create a safe environment for collaboration***
- ***Build group consensus about the nature and functions of collaboration***
- ***Honor diversity***
- ***Encourage diversity of opinion***
- ***Resolve conflicts***
- ***Encourage participation***
- ***Make sure that people are heard***
- ***Seek consensus about group decisions***
- ***Promote compromise***
- ***Facilitate problem-solving***
- Channel conflict toward positive outcomes
- Seek common ground in the face of conflict
- ***Balance group needs with individual needs***

Engage in continual personal growth

- ***Practice reflection***

---

* See the WIDS website at: http://www.wids.org/

- ***Examine the ethical framework in which you make decisions and interact with others***
- ***Model collaborative leadership core values of courage, compassions, continuous learning, and service***
- Regulate one's use of power and control
- ***Analyze one's learning styles***
- ***Explore one's spectrum of multiple intelligences***
- ***Inventory your internal and external assets***
- ***Draft a personal growth action plan***
- ***Develop capacity to act in the face of risk and uncertainty***
- ***Develop capacity for long-term vision***

Build communication networks

- ***Encourage open communication***
- ***Check for understanding***
- ***Listen actively***
- ***Promote active listening***
- Select communication channels
- Create opportunities for face-to-face communication
- ***Attend to non-verbal cues***
- Develop linkages with outside resources
- Communicate organizational identity
- Employ communication technologies

Nurture a positive group culture

- ***Facilitate the definition of the group's social contract (ground rules)***
- ***Foster a climate of continuous learning***
- Formulate group values, mission, and vision
- Model group values, mission, and vision
- ***Discourage groupthink***
- Critique group norms
- ***Seek diverse ideas and points of view***
- ***Support risk-taking***

Facilitate group processes

- Enable members to perform varied group roles
- Intervene to enhance group effectiveness
- ***Facilitate collaborative decision-making***
- ***Foster creativity***
- Facilitate negotiation
- ***Facilitate brainstorming***
- Initiate collaborative dialogue
- ***Choose the appropriate facilitation technique***

Organize resources

- Recruit members
- Identify talents of group members
- Obtain the expertise required to address the goals or problems
- Manage internal/external contact information

- Manage material resources
- Obtain financial resources
- Document financial transactions
- Manage budgets
- Manage risk

Accept responsibility for goal achievement

- Translate mission, philosophy, and ideas into goals
- Assess strengths, weaknesses, opportunities, and threats to goal attainment
- ***Set measurable outcomes***
- Develop an action plan for achieving outcomes
- Obtain group member commitment to complete action plan tasks
- Monitor action plan implementation
- Coach group members throughout action plan implementation
- Document goal accomplishment
- Celebrate achievements
- Analyze the impact of goal achievement

It is the hope of the Wisconsin Leadership Institute and the authors that you will find this curriculum not only informative and useful, but fun. It is well known, both anecdotally and empirically, that enjoyment increases the likelihood that actions will persist. This, by itself, will increase the chance that learning will be used and retained.

In the end, leadership should not be perceived as a burden, but as a way of being with others and the world. We invite comments, suggestions, questions, and criticism of this work. Any and all feedback is welcome. Thank you for joining this growing community of collaborative leaders.

UNIT 1 outline

What is Collaboration?

To excel as a collaborative leader means that one understands the nature of collaboration. For example, what is the difference between collaboration and cooperation? Although subtle, the differences between the two can mean the difference between following the crowd and being a voice in the crowd. It can mean saying nothing when a wrong is being committed, or being a voice of reason. And, it can mean the difference between doing things the way they have always been done, and making something happen that is new and exciting.

This unit explores the nature of collaboration and provides skills that help collaboration thrive. Because collaborative leadership takes place within the context of relationships, we start with developing group cohesion. Once people within a group have had an opportunity to get to know each other, it is then possible to look deeper to recognize (and appreciate) the diversity within the group. After delving into the similarities and differences between cooperation and collaboration, we then proceed to skills that allow collaboration to grow and deal with some of the barriers that get in the way of collaboration. These skills include: communication, decision making, conflict resolution and goal setting. The unit ends with a project that allows students to apply the concepts and skills learned through the unit.

A. Enduring Understandings

- Collaboration takes place within the context of relationships.
- Creating an environment that is safe and trusting encourages people to collaborate.
- Working collaboratively is a continual process that takes skill and practice.

B. Essential Questions

- What is collaboration?
- What are the similarities and differences between cooperation and collaboration?
- How do people come together to create an environment that supports collaboration?
- What are some useful skills to facilitate collaboration within a group?
- How do groups of people honor diversity so that it strengthens, rather than weakens, a group process?

C. Key Knowledge and Skills

Participants will know:

- How to "break the ice" with new groups.
- What it means to build consensus.
- That the diversity within a group can be a source of strength.
- What it means to be an ally.
- Some strategies to encourage participation within a group.

Participants will be able to:

- Articulate a definition of collaboration.
- Listen actively and attend to non-verbal cues.
- Create a social contract when working in groups.
- Use brainstorming as a decision-making tool.
- Reflect upon their learning about collaboration.
- Identify a variety of conflict-resolution tools and skills.

D. Unit Activities/Performance Tasks: (Experiential Lessons)

- Develop group cohesion.
- Recognize the diversity within the group.
- Differentiate between cooperation and collaboration.
- Use a variety of skills and tools for collaboration.
  - Communication Strategies
  - Decision Making Strategies
  - Conflict Resolution Strategies
  - Goal Setting Strategies
- Apply collaborative skills.

E. Varied Classroom Assessments and Rubrics

- Journal assignments.
- Chart qualities and attributes of productive vs. non-productive groups.
- Create a social contract for the class.
- Research project on someone who has gained or lost (or both) due to their group identity (portfolio).
- Create and post a class chart of ways to honor diversity in class and in life.
- Complete a Pi chart for cooperation.
- Write group and personal definitions of cooperation and collaboration.
- Create a class puzzle about collaboration.
- Read vignettes to determine where cooperation and collaboration are taking place.
- Create a rubric with some of the collaboration concepts.
- Create a list of skills that are helpful for productive group work.
- Group feelings role plays to illustrate non-verbal communication.
- Your personal network brainstorm.
- Return to *Shackleton* and *Into Thin Air* articles to find examples of decision-making.
- Identify possible conflicts through the Conflicting Needs activity.
- Identify conflict de-escalators through small group activity.
- Create before-and-after skits to show how conflict can be handled poorly and well.
- Personal SMART goal for teacher rating.
- Final group project for teacher rating.
- Portfolio completion chart.
- Portfolio evaluation.

F. Student Self-Assessment

- Journal assignments.
- Portfolio completion.
- Multiple Intelligence Inventory.
- Modifying one's definition of collaboration.

- Self assessment of collaborative skills with class-created rubric.
- Self assessment of active listening skills.
- Tracking of personal SMART Goal.
- Final group project rating.
- Portfolio evaluation.

G. Peer Feedback

- Definition of collaboration.
- 1-1 interview activity listening observation sheet.
- Non-verbal cues role play creation and evaluation.
- Before-and-after skits for conflict resolution strategies.
- SMART goal assessment.
- Final group project rating.
- Portfolio evaluation.

H. Logistics

- Time Frame: 28 to 52+ hours
  The time frames presented here are intended only as a guide. Each teacher has his or her own style, and every class has its own personality and needs. The amount of time invested in each part of the lesson is dependent upon choices the teacher makes about which activities to present, what is assigned for homework or done in class, richness of the discussion, etc.
- Materials to gather (other than copies of handouts): Please see each activity for intended use.
  - Deck of playing cards
  - 2 long ropes or tape
  - 3 x 5 note cards
  - Eye coverings: aka blindfolds (optional)
  - Posterboard, flip chart paper, and markers
  - 3-dimensional shapes of different colors. Make your own out of wood or go to a learning shop and buy a container of those large plastic buttons that come in geometric shapes.
  - One-quarter inch flat metal washer and a fifteen-inch string for each person.
  - Boundary markers (e.g., cones, polyspot markers, both may be available from the physical education department)
  - Channels made with 1/2-inch PVC pipe or corner moulding, ball bearing or marble, tin can or cup
  - Soft throwable objects (e.g., stuffed animals, wadded up pieces of paper): 1/person
  - Bandana for each group of 3 or 4, a cup for each group, lots of dry beans, broom
  - A large tarp with a 10 x 10 grid drawn on it with permanent marker (a grid taped to the floor can work, though it is obviously less portable)
  - Ball of yarn, chalkboard/whiteboard and markers, or butcher paper and markers
  - The book *Sneetches* by Dr. Seuss
  - Journals
  - Writing utensils
  - Legos®
  - Beach ball
  - Stopwatch

relevant standards

What is Collaboration?

Mid-continent Research for Education and Learning (McRel) standards* were chosen as an aggregate example of standards that are relevant to learning collaborative leadership within this curriculum. In 1990, McRel began a systematic collection of "noteworthy national and state curriculum documents in all subject areas." Their standards project is designed "to address the major issues surrounding content standards, provide a model for their identification, and apply this model in order to identify standards and benchmarks in the subject areas." (McRel, 2007) They serve as a useful comparison to the standards you use.

The standards and indicators as they are presented here are identified because they are explicitly addressed in this curriculum. There are other standards that can be attended to indirectly during the course of the class and the interactions that arise during activities and discussion.

BEHAVIORAL STUDIES

Standard 1: Understands that group and cultural influences contribute to human development, identity, and behavior.

1. Understands that cultural beliefs strongly influence the values and behavior of the people who grow up in the culture, often without their being fully aware of it, and that people have different responses to these influences.
4. Understands that people often take differences (e.g., in speech, dress, behavior, physical features) to be signs of social class.
6. Understands that heredity, culture, and personal experience interact in shaping human behavior, and that the relative importance of these influences is not clear in most circumstances.
7. Understands that family, gender, ethnicity, nationality, institutional affiliations, socioeconomic status, and other group and cultural influences contribute to the shaping of a person's identity.

Standard 2: Understands various meanings of social group, general implications of group membership, and different ways that groups function.

1. Understands that while a group may act, hold beliefs, and/or present itself as a cohesive whole, individual members may hold widely varying beliefs, so the behavior of a group may not be predictable from an understanding of each of its members.
5. Understands that social groups may have patterns of behavior, values, beliefs, and attitudes that can help or hinder cross-cultural understanding.

Standard 3: Understands that interactions among learning, inheritance, and physical development affect human behavior.

1. Understands that differences in the behavior of individuals arise from the interaction of heredity and experience.

---

3. Understands that expectations, moods, and prior experiences of human beings can affect how they interpret new perceptions or ideas.

Standard 4: Understands conflict, cooperation, and interdependence among individuals, groups, and institutions.

3. Understands that conflicts are especially difficult to resolve in situations in which there are few choices and little room for compromise.
2. Understands that intergroup conflict does not necessarily end when one segment of society gets a decision in its favor because the "losers" then may work even harder to reverse, modify, or circumvent the change.

CIVICS

What are the Basic Values and Principals of American Democracy?

Standard 11: Understands the role of diversity in American life and the importance of shared values, political beliefs, and civic beliefs in an increasingly diverse American society.

1. Knows how the racial, religious, socioeconomic, regional, ethnic, and linguistic diversity of American society has influenced American politics through time.
2. Knows different viewpoints regarding the role and value of diversity in American life.
3. Knows examples of conflicts stemming from diversity, and understands how some conflicts have been managed and why some of them have not yet been successfully resolved.

LANGUAGE ARTS: WRITING

Standard 1: Uses the general skills and strategies of the writing process.

11. Writes reflective compositions (e.g., uses personal experience as a basis for reflection on some aspect of life, draws abstract comparisons between specific incidents and abstract concepts, maintains a balance between describing incidents and relating them to more general abstract ideas that illustrate personal beliefs, moves from specific examples to generalizations about life).

Standard 2: Uses the stylistic and rhetorical aspects of writing.

6. Organizes ideas to achieve cohesion in writing.

Standard 4: Gathers and uses information for research purposes.

2. Uses a variety of print and electronic sources to gather information for research topics (e.g., news sources such as magazines, radio, television, newspapers; government publications; microfiche; telephone information services; databases; field studies; speeches; technical documents; periodicals; Internet).

LANGUAGE ARTS: READING

Standard 1: Uses the general skills and strategies of the reading process.

1. Uses context to understand figurative, idiomatic, and technical meanings of terms.
6. Understands the philosophical assumptions and basic beliefs underlying an author's work (e.g., point of view, attitude, and values conveyed by specific language; clarity and consistency of political assumptions).

Standard 7: Uses reading skills and strategies to understand and interpret a variety of informational texts.

3. Summarizes and paraphrases complex, implicit hierarchic structures in informational texts, including the relationships among the concepts and details in those structures.

4. Uses a variety of criteria to evaluate the clarity and accuracy of information (e.g., author's bias, use of persuasive strategies, consistency, clarity of purpose, effectiveness of organizational pattern, logic of arguments, reasoning, expertise of author, propaganda techniques, authenticity, appeal to friendly or hostile audience, faulty modes of persuasion).
5. Uses text features and elements to support inferences and generalizations about information (e.g., vocabulary, structure, evidence, expository structure, format, use of language, arguments used).

LANGUAGE ARTS: Listening and Speaking

Standard 8: Uses listening and speaking strategies for different purposes.

3. Uses a variety of strategies to enhance listening comprehension (e.g., focuses attention on message, monitors message for clarity and understanding, asks relevant questions, provides verbal and nonverbal feedback, notes cues such as change of pace or particular words that indicate a new point is about to be made; uses abbreviation system to record information quickly; selects and organizes essential information).

LIFE SKILLS: Life Work

Standard 7: Displays reliability and a basic work ethic.

1. Understands the concept of reliability (e.g., completing tasks on time; maintaining regular attendance; carrying out assigned tasks; being punctual).
5. Develops good work habits (e.g., keeping an effective work station; organizing job responsibilities).

LIFE SKILLS: Self-Regulation

Standard 1: Sets and manages goals.

1. Sets explicit long-term goals and shorter range subgoals.

Standard 2: Performs self-appraisal.

1. Knows a variety of learning styles and preferred learning styles.
6. Knows personal strengths and weaknesses and techniques for overcoming weaknesses.

Standard 5: Maintains a healthy self-concept.

2. Uses techniques to offset the negative effects of mistakes (e.g., sees mistakes as learning opportunities).
8. Understands that everyone makes mistakes, and that mistakes are a natural consequence of living and of limited resources.

LIFE SKILLS: Thinking and Reasoning

Standard 3: Effectively uses mental processes based on identifying similarities and differences.

1. Uses a comparison table to compare multiple items on multiple abstract characteristics.
2. Identifies the qualitative and quantitative traits (other than frequency and obvious importance) that can be used to order and classify items.

LIFE SKILLS: Working with Others

Standard 1: Contributes to the overall effort of a group.

1. Knows the behaviors and skills that contribute to team effectiveness.
2. Works cooperatively within a group to complete tasks, achieve goals, and solve problems.
4. Demonstrates respect for others' rights, feelings, and points of view in a group.
5. Identifies and uses the individual strengths and interests of others to accomplish team goals.

6. Identifies causes of conflict in a group and works cooperatively with others to deal with conflict through negotiation, compromise, and consensus.
7. Helps the group establish goals, taking personal responsibility for accomplishing such goals.
8. Evaluates the overall progress of a group toward a goal.
9. Contributes to the development of a supportive climate in groups.
10. Actively listens to the ideas of others and asks clarifying questions.

Standard 2: Uses conflict-resolution techniques.

1. Communicates ideas in a manner that does not irritate others.
6. Determines the causes and potential sources of conflicts.
7. Determines the seriousness of conflicts, and identifies explicit strategies to deal with conflict depending on its nature and seriousness.

Standard 3: Works well with diverse individuals and in diverse situations.

1. Works well with the opposite gender, of differing abilities, and from different age groups.
2. Works well with different ethnic groups, of different religious orientations, and of cultures different from their own.

Standard 4: Displays effective interpersonal communication skills.

1. Demonstrates appropriate behaviors for relating well with others (e.g., empathy, caring, respect, helping, friendliness, politeness).
3. Knows strategies to effectively communicate in a variety of settings (e.g., selects appropriate strategy for audience and situation).
5. Uses nonverbal communication, such as eye contact, body position, and gestures effectively.
6. Demonstrates attentive listening by clarifying messages received (e.g., paraphrasing, questioning).
8. Attends to both verbal and nonverbal messages.
11. Demonstrates sensitivity to cultural diversity (e.g., personal space, use of eye contact, gestures, bias-free language).
13. Acknowledges the strengths and achievements of others.

Standard 5: Applies basic trouble-shooting and problem-solving techniques.

7. Evaluates the effectiveness of problem-solving techniques.
8. Reframes problems when alternative solutions are exhausted.
10. Evaluates the feasibility of various solutions to problems; recommends and defends a solution.

Standard 6: Applies decision-making techniques.

5. Evaluates major factors (e.g., personal priorities, environmental conditions, peer groups) that influence personal decisions.
6. Analyzes the impact of decisions on self and others and takes responsibility for consequences and outcomes of decisions.

Physical Education

Standard 5: Understands the social and personal responsibility associated with participation in physical activity.

3. Understands how participation in physical activity fosters awareness of diversity (e.g., cultural, ethnic, gender, physical).
4. Includes persons of diverse backgrounds and abilities in physical activity.

Lesson 1.A Overview: Develop Group Cohesion

Focus: Create a safe environment for collaboration
Facilitate the definition of the group's social contract (ground rules)

1

Create some basic ground rules based on the idea of safety and respect.

10 - 20 min

2

Choose from the Ice-breaker, Deinhibitizer, and Trust activities.

And

Use the processing questions included with each activity to explore issues of put ups/put downs, mixing, and trustworthiness.

1.5 - 4 hr

3 (portfolio)

Have students keep track of all activities done for their portfolio.

5 min to present

4 (assessment)

Journal assignment.

10 - 15 min

5

Read the synopsis of Shackleton's voyage.

20 - 30 min

Or

Watch the film: *Shackleton's Voyage of Endurance.*

2 hr

Or

Read the transcript of the film: *Shackleton's Voyage of Endurance.*

1.5 - 2 hr

6

Read the synopsis *Into Thin Air: A Personal Account of the Mount Everest Disaster* by Jon Krakauer

25 - 35 min

Or

Read the entire article from *Outside Magazine.*

1 - 1.5 hr

7

Hold small group discussions about the two events.

20 - 30 min

8 (assessment)

Create a chart of qualities and attributes of productive and non-productive groups.

15 - 20 min

9 (assessment)

Create a social contract for the class.

20 - 30 min

Time Frame: 4 - 9 hours

LESSON 1.A SEQUENCE: DEVELOP GROUP COHESION

Create some basic ground rules based on the idea of safety and respect.

Ask students what they need in order to feel safe to do some activities and write their suggestions on the board (e.g. respect, be listened to, non-judgmental, etc.). This is a way to begin to create ground rules and establish norms for a healthy functioning group. Periodically check in to see if people are living up to the ideals that have been identified. If so, note that. If not, note what needs to change to put these ideals into action. Post these in the room.

At the end of this lesson, the class will create a more permanent social contract for the class.

Choose from the following Icebreaker, Deinhibitizer, and Trust activities. Use the processing questions included with each activity to explore issues of put ups/put downs, mixing, and trustworthiness.

These activities are designed to help group members get to know each other better and begin to establish a foundation of support and trust. Make sure students are aware that creating a sense of cohesion and community is **essential** for the development of collaboration and that it is being done **intentionally**.

It is important to remember that group cohesion takes time to develop, and must be undertaken with care and gentleness. If someone is quiet, he or she can be invited to share, but should always have the right to pass. As people become more comfortable, they will open up when they are ready.

Do as many activities as time allows. Stay with the icebreaker activities until students seem ready to move on to deinhibitizers. Some indicators are: a decrease (or, ideally, elimination) of put downs and the use of sarcastic humor, a willingness to work with anyone in the class, and an ability to listen to others.

Students will indicate a readiness to move from deinhibitizers to trust activities when you notice they are laughing with (rather than at) each other, don't mind acting silly around their classmates, and show a true sense of caring toward others in the class.

Leadership, especially collaborative leadership, takes place within the context of relationships. It is important for individuals to acquire the skills to build and nurture relationships with others. Each of the following are skills that are helpful for fostering relationships:

> **Put ups/Put downs:** Encouraging rather than demeaning self and others.
> **Mixing:** Being able to work with everyone in the group.
> **Trustworthiness:** Proving that one is worthy of the trust of another by showing support and caring.

After the activities, touch base about what happened and make connections to how people work together, and will be working together during the class. It is this "processing" that helps students make connections between what they are experiencing and how it can be used.

The questions provided are samples, and can be used to help trigger a conversation. Please substitute or add your own questions and comments to address the needs of your class. The ultimate goal is for the students to intentionally build positive relationships in order to work together collaboratively.

Icebreaker Activities

Provide introductions

Get-to-know-you experiences

Include name games

Help people relax and have fun

Help people feel included

Stay with the ice breaker activities until students seem ready to move on to deinhibitizers. Some indicators are: a decrease (or, ideally, elimination) of put downs and the use of sarcastic humor, a willingness to work with anyone in the class, and an ability to listen to others.

activity

Cards

Focus: Working together, mixing, mild problem solving
Materials: Decks of cards
Time: 10 - 20 minutes
Sequence: Ice Breaker
Sources: Laurie Frank learned this activity from Chris Cavert and Jim Cain

Suggested Procedure

1. Sort the cards so there are enough for your large group to get into smaller groups. For example, if you have a group of 30, use the aces through fives, having six of each card (6 aces, 6 two's, etc.) with a mixture of suits.
2. **Part 1:** Hand out the cards. It's okay for participants to look at them.
3. Have each person find a partner and tell him/her, then have partners tell each other as many things about themselves as the number on their card.
4. Trade cards and find a new partner.
5. **Part 2:** Have them continue to trade cards with 3 to 5 other people **without looking** at the cards.
6. Still without looking, have them place the card face out on their forehead and, without talking, get into groups (according to color, suit, number).
7. Once in their groups, have them do an activity together or discuss something.

Sample Processing Questions

- How did you feel about sharing information with another person?
- How did you choose the partners with which to share?
- How did you find the group of people with similar cards if you couldn't see the cards and couldn't talk?
- What kind of communication did you use?
- Did you tend to show other people where they might go, or did you wait for others to give you instructions?
- How did people help each other to accomplish this task?

Facilitation Notes

During the first part of the activity, be alert for group members who are holding back. Help them find a partner or be their partner. During the second part of the activity, groups sometimes hesitate thinking there is no way they can possibly find others without being able to look or speak. They quickly pick up on other ways to communicate and how they need to help each other in order to accomplish the task.

activity

Group Bingo

Focus: Mixing with others in the group, learning about others, perspective taking
Materials: Pencils and pre-made group bingo cards (see p. 24)
Time: 10 - 15 minutes
Sequence: Ice Breaker
Sources: See "Human Bingo" in *Adventures in Peacemaking* by Kreidler & Furlong, and *Journey Toward the Caring Classroom* by Frank.

Suggested Procedure

1. Each person gets a group bingo card.
2. The task is to get as many different signatures as possible in the amount of time allotted.

Sample Processing Questions

- Did you learn anything about others in the group? What?
- Does anyone have a personal story to tell about any of the things on the card?
- What else do you want to know about others in the class?

Facilitation Notes

It is important to take some time after the activity to compare notes about people in the class. Ask how many left-handers you have in the room, or how many people have broken a bone. Encourage individuals to share stories, so that everyone has the opportunity to learn more about each other.

Rather than using the pre-made bingo card, have the class make their own. Split into smaller groups and come up with three to five things they want to know about others in the class. Then make a bingo card that includes the students' questions.

activity

Name High Five

Focus: Name reminder
Materials: None
Time: 5 - 10 minutes
Sequence: Ice Breaker
Sources: Laurie Frank learned this activity from Steve Butler.

Suggested Procedure

1. Clear the desks or tables away and stand in a circle.
2. Go around the circle and have everyone clearly say their first name so that all can hear.
3. You step out into the middle of the circle and call out one person's name.
4. That person joins you in the middle and you do a high five.
5. You, then, go to that person' place. He calls out someone else's name, who joins him in the circle and they high five.
6. He goes to her place, and she calls out someone else.
7. This continues for a bit. Then you can either speed things up or have two or more people in the middle calling names.

Group Bingo

Someone who has a library card.	Someone who has different color hair than you.	Someone who has waded or been swimming in an ocean.	Someone who is left-handed.	Someone who likes similar movies as you.
Someone who is the same birth order as you (first/only born, middle, last born).	Someone who has been to a state outside the one in which you live.	Someone who has been outside of the United States.	Someone who can speak more than one language.	Someone who has a pet other than a cat or dog.
Someone who has lived in an apartment.	Someone who enjoys knitting or crocheting.	FREE SPACE	Someone who plays a team sport.	Someone who had their nails done.
Someone who was born on an even numbered birth date.	Someone who has more than one pet.	Someone who likes to eat broccoli.	Someone who knows ASL (American Sign Language)	Someone who celebrates different holidays than you.
Someone who has been to the top of a mountain.	Someone who has broken a bone.	Someone who has done community service.	Someone who belongs to a club or group outside of school.	Someone who likes to fish.

Sample Processing Questions

- Who can name everyone here? Give it a try.
- Mix up and see if anyone wants to try to name everyone in the group.

Facilitation Notes

This activity is very straightforward—its sole purpose is to help people remember each others' names. Just try to keep it moving by speeding it up or adding more people in the middle. Also keep an eye out in case there is one person who is not being chosen by anyone.

activity

Paired Activities

Focus: Learning about others, Mixing with others in the group
Materials: None
Time: 10 - 20 minutes
Sequence: Ice Breaker
Sources: Laurie Frank learned these from a variety of sources including: Pete Albert, Dick Jensen and Jim Dunn. See Journey Toward the Caring Classroom by Frank and *More New Games* (Last Detail) by Fluegelman.

Suggested Procedure

1. Clear the desks or tables away and stand in a circle. Ask people to get a partner (it doesn't matter who because people will be with different partners later).
2. Have the partners look at each other. Name an attribute (shorter hair, darker eyes, the most blue on, etc.) and have them separate themselves accordingly, e.g., the person with shorter hair stands inside the circle facing their longer-haired partner who stand outside the circle. There are now two circles, one inside the other, with the inside circle facing the outer one.
3. Give the pairs an activity to do (they are listed below).
4. After the activity, have partners discuss something about themselves. Possible discussion topics: your family, your favorite place in the world, your favorite food, a hope or goal in your life, something you hope to learn someday, etc.
5. After the activity and discussion, have one of the circles move a couple people to the right or left. Each person then has a new partner, and is ready for a new activity and discussion topic.

Activities

Last Detail: Everyone looks at their partner. On a signal, everyone turns around and changes three things on their clothing (turn collars, switch shoes, take out earrings, etc.) When each pair is ready, they turn around and try to guess the three things that were changed.

Tie Your Shoe: Each pair should have at least one person with tied shoes, or the ability to borrow a shoe from someone else. The shoes are untied, and the task is to retie the shoes. The problem is that each person in the pair can use only one hand.

Me Switch: Teach the students three motions. It doesn't really matter what they are, but here is an example: Both hands above eyes, both arms crossed over the chest, and one hand touching the other elbow. Then have each pair designate a person who is "it." This person counts to three, at which point both people simultaneously go into one of the motions. The person who is "it" wants the other person to do the same thing that he is doing. If she does, she is now "it," and does the counting. If she doesn't do the same thing, then he remains "it" and counts again.

Macro Rock/Paper/Scissors: Make sure everyone knows the rules to Rock/Paper/Scissors: Rock = closed fist, Paper = open hand, and scissors = fingers in a V like a peace sign. On a signal, the partners do one of the three signals to see who wins. The winning combinations are: Rock beats scissors because it crushes the scissors, scissors beats paper because it cuts the paper, and paper beats rock because it covers the rock. It is also possible to tie. For the macro version, substitute these motions: Rock = crouching down with hands over head, Paper = standing with arms at sides, Scissors = standing with arms over head in a V. Have partners face back-to-back. Each person must take a step forward to avoid hitting the other. They count to three while jumping up and down, spin around and go into one of the motions. Same rules apply. (Thanks to Dick Jensen and others at Toki Middle School for this variation).

High Fives: With their partners, create three new high fives. Tell people to be creative, and that it's okay to "cheat" by looking around at others for ideas. Once done, ask each pair to choose their favorite to share with the group.

Celebration: Partners create some way to celebrate—high fives, a dance, a cheer, etc. Then, throughout the day, when someone yells "celebrate," they must find their partner and celebrate with them (from Jim Dunn).

Sample Processing Questions

- How was it to work with so many different people?
- Why might it be important to be able to work with everyone in this group?
- Is it easy or hard for you to approach someone new?

Facilitation Notes

As you can see, these activities have no deeper underlying meaning in and of themselves. They are simply fun things to do with another person and are generally non-threatening. Some groups need to hear that there is no point to the actual activity, but that there is a purpose to the exercise as a whole.

Depending upon the group, it may be necessary to help them get initial partners, especially if you are aware of triads or students who are regularly excluded. Try having them line up in alphabetical order and folding the group in half. Other strategies include: Find someone who was born in a different season than you, or find someone who has the same size thumb (or different) than you, or find someone whose eyes are the same color as yours. It is also possible to get partners at random by using a deck of cards, with the number of cards equaling the number in the class, and each card having a partner (such as two kings, two aces, etc.).

Deinhibitizer Activities

Expect more of people

Barrier Breaking

Help people begin to make choices about risk taking

Allow people to laugh with, and not at, each other

Allow people to act silly, be put on the spot, and appropriately touch others

Students will indicate a readiness to move from deinhibitizers to trust activities when you notice that they are laughing with (rather than at) each other, don't mind acting silly around their classmates, and show a true sense of caring toward others in the class.

activity

Night At the Improv

Focus: Spontaneity, being put on the spot, active listening
Materials: None
Time: 10 - 20 minutes
Sequence: Deinhibitizer
Sources: See "Story Circle" in *Adventure Education for the Classroom Community* by Frank and Panico, *Journey Toward the Caring Classroom* by Frank, and "Theater Sports" in *Teamwork and Teamplay* by Cain & Jolliff.

Suggested Procedure

1. Clear a space and have people get into pairs.
2. **Phase I:** With a partner, show how you can create a sentence by having each person add every other word. The trick is to listen to what your partner is saying, rather than what you think they will say. For example:
   You: "Once" Partner: "there"
   You: "was" Partner: "a"
   You: "yellow" Partner: "jacket"
   You: "that" Partner: "flew"
   You: "up" Partner: "under"
   You: "your" Partner: "shirt"
   You: "period"
3. Have partners practice for a while.
4. **Phase II:** Now get the group into a circle.
5. Create sentences with the whole group by having each person contribute a word around the circle. Do this for a while.
6. **Phase III:** Create a whole story. Designate a stage area.
7. Have someone give a sentence that is the beginning of a story. For example, they might say: "Once upon a time," or "Call me Ishmael," or "The whole thing was a big mess." This person stands on the stage area on one end.
8. Then have someone give the end of the story. They might say, "And they lived miserably ever after," or "They rode into the sunset," or "The whole thing was a big mess." This person stands on the other end of the stage area. Have them say the story so far, beginning to end.
9. Have someone give the middle of the story. They might say, "We thought the truth was out there," or "He fell asleep for a long, long time," or "The whole thing was a big mess." This person stands between the beginning and the end. Say the whole story, as is.

10. Now, have people begin putting themselves into the story with their own lines, one or two people each time. Then, repeat the story as it stands.
11. Finally, the whole class is in the line, and the entire story can be recited with emotion.

Sample Processing Questions

- How did you like being put on the spot to come up with a word or phrase so quickly?
- During the story, did you wait until the story unfolded or did you jump in right away? Why did you use that particular strategy?
- Was this difficult or easy for you? Why do you think so?
- Did you feel supported by your classmates? Why or why not?

Facilitation Notes

If the students are ready to be spontaneous, this can be a wonderfully fun activity. Even if they aren't feeling very spontaneous at the beginning, try doing this periodically throughout the year. It is good practice for listening, and the creativity level rises each time it is attempted.

During the story, some people stand back for a while to reflect more on the story, while others jump in right away. This can open the door to a discussion about people who are more action oriented versus those who prefer to reflect before taking action. If necessary, remind people that punctuation is needed, or they can be "the end." If someone cannot think of a place to go, that's fine. It's always good to have an audience. For ground rules, remind people to keep it "G" rated, and not to use the names of classmates, teachers, neighbors, etc.

activity

Cross the Line

Focus: Creativity, cooperation, taking turns, perspective taking, communication, choices, leadership
Materials: Long rope
Time: 15 - 20 minutes
Sequence: Deinhibitizer
Sources: See *Diversity in Action* by Chappelle and Bigman.

Suggested Procedure

1. Lay the rope out in a circle and stand around it.
2. The object is for everyone to cross the line differently from anyone else.
3. Go around the circle with each person taking a turn to cross the line.
4. Next have everyone pair up.
5. Give them time to come up with a way to cross the line differently from any other pair.
6. Go around and have every pair show how they will cross the line.
7. Get into groups of 4 and do the same, then groups of 8.
8. Eventually the whole group will cross the line together.

Sample Processing Questions

- How did your communication and decision making change as you got into larger groups?
- Which did you find easier—crossing the line alone, in pairs, or in groups? Why?
- Did you find that you were willing to go along with an idea, or did you have an idea that you pushed to be used by the group?

- What role do you generally take in a group decision-making process? Do you hang back, or are you right in the thick of things?
- Was it easier or more difficult to come up with ideas alone—with others?

Facilitation Notes

This is a great activity to use when exploring how people react in a group decision-making process. Some people are naturally shyer than others. They find it difficult to do something alone and feel more comfortable in a larger group because it's easier to be anonymous. Others like to show off by themselves.

activity

HANDSHAKES

Focus: Mixing, meeting new people
Materials: None
Time: 10 - 15 minutes
Sequence: Deinhibitizer
Sources: Laurie Frank learned this activity from Jim Cain.

Suggested Procedure

1. Have everyone get a partner.
2. Teach them the fish handshake: Start like a regular handshake, then move up to the forearms and gently slap your partner's forearm as they do the same.
3. Have everyone get a new partner and teach the farmer handshake: One person interlocks his fingers and turns his hands upside down, letting his thumbs dangle. The other person squeezes her partner's thumbs (thus simulating the milking of a cow or goat).
4. Then have everyone get a new partner and teach them the logger handshake: One person makes a fist and sticks the thumb up. The other person grabs that thumb and does the same. The first person grabs the second thumb. Finally, the second person grabs the last thumb. Together they make a sawing motion while saying each other's name.
5. Have everyone get another partner and make up their own handshake. After all of this, have them find their handshake partners to do their handshakes one more time.

Sample Processing Questions

- Did it matter who your partner was? Why or why not?
- Why might it be important to be able to work with everyone in this group?
- Did you learn anything about any of your handshake partners?

Facilitation Notes

As a variation, have partners share something about themselves with each of their handshake partners. This is also a "handy" way to get partners later. Simply ask students to "find your fish handshake partner." It takes but a minute for them to find their partners, do their handshake, and be ready.

Explore risk taking

Look at idea of being trustworthy

Help people decide when, how and where to trust others

Give people the opportunity to trust and be trusted

activity

Yurt Circle

Focus: Responsibility, trustworthiness, empathy, cooperation
Materials: None
Time: 15 - 20 minutes
Sequence: Trust/Support
Sources: See *Adventure Education for the Classroom Community* by Frank & Panico, *Adventures in Peacemaking* by Kreidler & Furlong, *Changing the Message* by Albin, and *Journey Toward the Caring Classroom* by Frank, among others.

Suggested Procedure

1. Find a space that is big enough for the whole group to stand in a shoulder-to-shoulder circle, and then have everyone take a step back.
2. There must be an even number of people (you can make up the difference).
3. Determine the state bird and state flower. In Wisconsin, it is the robin and wood violet.
4. In this case, every other person is labeled a robin, and the others (every other one as well) are wood violets.
5. The object of this activity is for all of the robins to lean one direction while all the wood violets are leaning the other direction. Determine which direction each group will lean.
6. When people lean, they should keep their bodies as stiff as possible, trying not to bend at the waist.
7. Ask everyone to hold hands so that they have a good grip. Remind everyone not to let go during the activity.
8. Count to three, and then have people slowly lean in their given direction.
9. Try this a couple of times, then switch leaning directions. When people get really good at this, they can start one direction, then switch to the other direction without starting over.
10. Note: An easier version of this is for every other person (i.e., all the robins) to turn around. When people join hands and lean, everyone leans back. This way all the robins lean toward the middle of the circle, and all the wood violets lean away from the circle. After you try this, then have everyone face in and try leaning in and out.

Sample Processing Questions

- What made this work?
- What would happen if someone let go?
- Was it easier for you to lean in or out? Why?
- How did we take care of each other and show responsibility toward each other during this activity?
- How might a community work in this way, where everyone is connected, even though it might not be obvious?
- How did you work with the people around you to make sure no one was pulling too hard? What adjustments did you make?

Facilitation Notes

It may be necessary to address the issues of holding and/or squeezing hands. First, determine if group members are ready to hold hands. Then ask them, or remind them, about trust and being careful with others. You might suggest that they holding wrists, which gives a better grip.

This activity shows how people are connected, even when they do not think they are. If one person moves too fast, or is not in sync, then everyone feels it. It usually takes a few tries to get the yurt circle to work well. Once people get it, then it seems almost easy.

activity

60 Second Speeches

Focus: Risk taking, emotional trust, empathy
Materials: 3" x 5" note cards and writing utensils
Time: 30 - 50 minutes
Sequence: Trust/Support
Sources: See *Adventure Education for the Classroom Community* by Frank & Panico, and *Journey Toward the Caring Classroom* by Frank.

Suggested Procedure

1. Everyone is given a set amount of time (5 to 15 minutes) to prepare a 60-second speech of their choosing.
2. Offer note cards for those who wish to take advantage of them.
3. Set up a forum for the speeches—auditorium-like, circle, etc.
4. Each student is then given 60 seconds for their speech, no more and no less.

Sample Processing Questions

- What was the most difficult part of this activity for you?
- How did the audience treat you? Were they helpful?
- What strategies did you use to get through this challenge?
- Did you consider this a risk or not? What do you think makes it risky for some and not so risky for others?
- What things do you find risky?
- Did you want to get this over with, or were you content to put it off as long as possible? How do you usually handle situations that are difficult or risky for you?

Facilitation Notes

Speaking in front of a group is one of the riskiest endeavors for people in our society. Take the time to acknowledge this fact, and then talk about what the class can do to make it easier for each speaker. Giving a speech is to risk embarrassment. How does one show their trustworthiness with another's feelings?

Since this activity can take some time, split it up over a few days. Have a few people give a speech at the end of a class period, or as a warm up to another activity. It is a great change-of-pace activity.

Trust Walk

Focus: Physical and emotional trust, trustworthiness, empathy, risk taking
Materials: Eye coverings (optional)
Time: 20 - 30 minutes
Sequence: Trust/Support
Sources: See *Journey Toward the Caring Classroom* by Frank, *Diversity in Action* by Chapelle & Bigman, *Games for Change* by Dodds & Prosser-Dodds.

Suggested Procedure

1. Depending upon age and maturity level of the students, this activity can be done inside or outside (see facilitation notes).
2. Have your students get into pairs, either by choice or by a random method.
3. Tell students they will take turns leading someone whose eyes are closed.
4. Go over all safety guidelines: Taking it slow, making sure both people can fit through a space, looking for obstacles, giving your partner as much information as possible, making sure they know what is around them.
5. Show the class different methods for leading someone who cannot see. They may wish to hold hands, hold the elbow of their guide, or have the guide hold their elbow. They may not wish to be touched at all, but opt for verbal directions only. The important thing is that they have this discussion. Each person must be given the opportunity to choose if or how to be touched.
6. Teach the "bumpers up" position (hands out in front) for the person whose eyes are closed.
7. Remind students that there are no secrets about what is in the area, and to feel free to describe what or who is around the person they are leading. Encourage them to take their partner to an object (tree, chair, etc.) and have them feel it. Later they can try to guess which one it was.
8. At any time either person in the partnership can stop the activity if they feel too much discomfort. They must, of course, communicate the decision to their partner before stopping.
9. Offer eye coverings for those that want them. Others can just close their eyes.
10. Set boundaries.
11. Allow enough time for each person to have a turn guiding and being led.

Sample Processing Questions

- What did your guide do to make you trust her/him? Be specific.
- Did you feel your guide took care of your safety? Why or why not?
- Did your guide do anything to make you nervous? What?
- Were you more comfortable being led or being the guide?
- When you were guiding, did you feel responsible for your partner? Why or why not?
- Did you feel the need to peek? What caused you to choose to look?
- How risky was this for you? What would have made it more/less risky?
- What do you think causes someone to be trustworthy, which helps build trust with others?

Facilitation Notes:

At first have your students try this activity in a large open area like a gym or empty lunchroom. It will give them an opportunity to understand all that is expected, and give you the opportunity to assess their readiness to accept the responsibilities that go along with building trust. Add obstacles in the large room, then try these activities in a smaller space, like your classroom, or outside where the terrain is uneven and less predictable.

It is also important to put some thought into how you want your students to pair up. In the beginning, it may be helpful to have your students pick their own partners, to ensure that they are comfortable with the person who is guiding them. As their comfort level increases, have them pair up in a more random fashion so that they have an opportunity to build trust with people they do not know (or even like) as well.

During the processing time students generally talk about how it helped to know that their partner was there by hearing their voice or feeling their hand on their shoulder. This is a good metaphor for communication. Good communication can give rise to a higher level of trust.

1.A sequence 3

Have students keep track of all activities done for their portfolio.

Have each student create a portfolio of his or her work from this class. This will allow them to refer back to earlier work for later projects. It will also be graded by themselves, their peers, and the teacher at the end of the course as a "final exam."

Journal assignment.

Journal Question: Describe your experience of getting to know the people in this group. What were your feelings when you came into this class on the first day? What are they now?

Read the synopsis of Shackleton's voyage
OR Watch the film: *Shackleton's Voyage of Endurance*[1]
OR Read the transcript of the film: *Shackleton's Voyage of Endurance*

(see bibliography on p. 245). This adventure shows how group work can succeed, even in the face of overwhelming odds.

Shackleton's Voyage of Endurance

The Shackleton Voyage to Antarctica lasted 21 months. Originally planned as a journey to cross Antarctica, it ended as a fight for survival. What helped these men survive an experience that had caused others to starve, go insane, and die?

It was the year 1914 when the *Endurance* left England with 28 men on board. The plan was to sail to the southernmost continent and, with a smaller crew of men with sleds and dogs, cross the 1800 miles of ice-covered land. Shackleton's past trips to the South Pole had ended short of their goal. As this transcontinental expedition had never been done, Shackleton would be immortalized as a true polar explorer.

Ernest Shackleton was charismatic and good at raising money. He chose his crew carefully and believed that character mattered more than experience. Everyone who signed up knew what he was getting into, as Shackleton was clear about the hazards even when advertising for the crew:

> "Men wanted for hazardous journey. Small wages. Bitter cold. Long months of complete darkness. Constant danger. Safe return doubtful. Honor and recognition in case of success. Ernest Shackleton." (Public Broadcasting System [PBS], 2002)[2]

He knew this expedition would try the mettle of every person on board, but he could not have predicted the colossal challenges that were in store.

After leaving England, the *Endurance* sailed to the tip of South America, where they headed south to South Georgia, an island off Antarctica. Shackleton was able to get information about the conditions at a whaling station there. The whalers were noticing that the ice was particularly bad that year. He learned that the Weddell Sea was known for ice that crushed ships. It was November 5th, summer in the southern hemisphere. If he waited much longer, the expedition would be in jeopardy because of the approaching Antarctic winter. He decided to depart.

As they sailed south, they noticed ice much further north than expected. Their wooden vessel was battered by big blocks of ice. They spent weeks hearing the ice hammer the *Endurance* and using valuable fuel (coal) to cut through it.

Mid-January brought them only a day's sail to the coast of Antarctica, but the ice was packing up. Even though they could see the open water, the ship strained to get through, and their fuel was being used too fast. Shackleton, then, made a critical decision to wait for the way to clear. Instead, the ice closed in around them, and the open water froze. They were only 30 miles from land.

> From the diaries, it would appear that they were all made by Shackleton to get on with the daily routine, which obviously had the effect of keeping morale up. And I think Shackleton himself, with his Irish background and ability to communicate and join in, made everybody feel that they were one. It was a team and not a "them and us" situation. Despite differences in rank and background, he insisted that everyone share equally in the burden of ship's chores (PBS, 2002).

In mid-February, open water was sighted. They tried to cut a way through using saws and picks. After working for two days without a break, they finally gave up as the water froze as quickly as they broke it up. Shackleton never showed defeat. Even as he realized that he would not achieve his goal of getting across Antarctica, he was calm when he told the crew that they would have to spend the winter on the ship. His task was to keep the morale of the crew high because he knew that the inhospitably long, dark, cold Antarctic winter lay ahead—a winter that could easily drive people crazy.

So, they spent the long dark hours following a routine and entertaining themselves with singing and shows. They lived together like a family. There was some conflict between crew members, though less than expected under the close, cramped, and harsh conditions.

As the *Endurance* drifted with the ice over 600 miles to the north they endured more frequent blizzards with winds up to 70 miles per hour. Ironically, this meant that Spring was coming, and the incessant noise meant that pressure was being applied to the ship by the ice moving around. They were stuck in the ice in the middle of the Weddell Sea, and their ship started to come apart.

Shackleton was calm as they watched the ship come apart and decided what to do. He knew that the real enemy was not the ice, but the morale of the crew. He decided to abandon ship on October 27th, 1915. The lifeboats were lowered to the ice, and they were stranded. The men, now, knew conclusively that the expedition was no longer an option.

It was now a matter of surviving and getting home. They were left with five tents and 18 sleeping bags. Rather than giving the sleeping bags to the officers, Shackleton had the crew draw lots, making sure he and some of those in authority got the blankets instead of the fur-lined sleeping bags. This meant that those with a lower rank had as much of a chance as those with higher rank at surviving. They were a team.

> Enmities can be sometimes extremely destructive to the harmony of an expedition—and so can alliances. So he [Shackleton] moved people around and noticed how people were getting on with each other or not getting on with each other. But it was all based on knowing his men. It's no good knowing theoretically how to handle people if you don't really notice what people are like. And he was extremely observant (PBS, 2002).

They worked toward getting to land by hauling the boats over the ice. Fifteen men pulled the lifeboats. After two days of pulling, they had gone less than two miles. Finally, they decided to wait and hope they would drift closer to land. This meant living on the ice.

A group salvaged what it could from the *Endurance* and, on November 21st, 1915, the crew watched as the *Endurance*, their "home," broke up and sank. They spent two-months living on the floating ice before trying a second time to reach land. One of the crew members (the carpenter) balked and refused to obey orders. Shackleton knew that he had to stop this quickly because, in a survival situation, unity is essential—a matter of life and death. He let the entire crew know that they could only survive as a team. He explained that he was not only the leader of the expedition but their boss, and he would make sure they got paid. The would-be mutineer backed down.

Even with everyone working together, they were only able to advance a mile per day. After seven days, they gave up. The effort was not lost, however, because the crew knew they had done all they could. Strangely, this helped morale. The only comfort they had now was food, even as the food supply was dropping. As the winter approached there would be less food, but Shackleton refused to stockpile food. He knew that this would be a blow to morale because it would prove that they had no hope. It was a choice between starvation and insanity, and he chose starvation. Eventually, they had to use the dogs for food.

They drifted for three months on the ice floe before it started to break apart. At this point, Shackleton decided to put the lifeboats in the water and head for Elephant Island. So in April, 1916, after 14 months of being locked in the ice, they dropped the boats in the treacherous water and made a break for land. For over three days they rowed while they shivered through icy water. They had no fresh water and almost no food. By the time they made it to Elephant Island, half of the men were showing signs of mental illness. They had made it to land just in time. It had been 16 months since they had touched land.

After eating their first hot meal in five days and getting some sleep, they could assess the situation. They were still far from civilization—how were they to get back? The nearest human beings were 400 miles away, but the winds did not blow that direction. This meant that they would have to go 800 miles to the whaling station on South Georgia Island. The only way there was by boat.

Everyone knew the dangers; no one had ever survived such a journey. Shackleton called the crew together and explained his plan. Of the many who agreed to go, he again chose carefully. He chose some people for their obvious skills: the navigator was an expert at sailing small boats, and the carpenter could keep the boat seaworthy. He also chose an experienced veteran of polar expeditions, a young man who proved his worth on the way to Elephant Island, and a man who was trouble on land, but tough at sea.

Even though the carpenter was the earlier would-be mutineer, Shackleton knew he needed him:

> The most important man onboard ship is not the captain or the navigator. It's the shipwright, because he's the man who keeps her afloat and makes her able to stand the fury of the ocean. He was one of the very few people alive, probably, who could have prepared her for the strains and buffeting of an oceanic voyage (PBS, 2002).

With only the meager parts on hand, he was able to make the boat seaworthy.

April 23rd saw the little lifeboat head toward the ocean with its small crew. Shackleton left his second-in-command in charge of the motley group staying behind. It is difficult to believe that they had any hope of being rescued, but what other options did they have?

The small boat and crew endured bad weather for most of the voyage. They had to depend on their navigator to get them where they wanted to go. Without him, they would have simply headed out to sea. Unfortunately, the weather made it difficult to get any bearings. The sun and stars were necessary to figure out where they were. After six days he was able to figure out that they had sailed 238 miles and were in the middle of the Southern Ocean.

As they headed into yet another storm that lasted for five days, sometimes riding 30-foot waves, the seasick men kept their spirits up with the help of their young crew mate. Although he didn't have any navigational skill, he was an unbridled optimist. Without him, they might have gotten overwhelmed by the enormity of what they were trying to do. Shackleton, too, made sure people got what they needed when they began to falter.

Finally they saw the first signs of land, but the cold, icy waters were not yet through with them. Hurricane-force winds drove them the wrong way. Their navigator, though, was able to somehow steer them to shore. It was nothing short of a miracle.

> What got the James Caird to South Georgia was a combination of luck and skill. The skill was Worsley's, this brilliant navigator, this wonderful small boat handler. But there is always the

> element of luck. And this is where we come to the great imponderable: Shackleton was a lucky man. He was lucky (PBS, 2002).

They had spent a total of 17 days at sea, and were amazed to be on solid ground.

One would think the story was over, but they had landed on the wrong side of the island. The little lifeboat could not handle the 150-mile trip, so they would have to do the unthinkable—walk over the mountains to the other side of the island. Even with mountain climbing equipment, this had never been done. This ragtag group with no equipment and little food would find it virtually impossible.

Nevertheless, Shackleton and two others from the group set out with a rope from the boat and screws put into their boots for traction. They scaled the mountains and crossed uncharted glaciers to reach their destination, marching for almost 26 hours straight. When they arrived at the whaling station, they were ghosts of their former selves such that the whalers did not recognize the men who had been there almost a year-and-a-half earlier.

The whalers were quick to mount a rescue mission, but 10 days of bad weather delayed the trip. Meanwhile, the men on Elephant Island were held together by Shackleton's second in command. Every day he made sure they followed a routine, and he kept talking about the impending rescue. Even as the days wore on, and people's health started to wane, he had people doing chores, singing, and writing in their diaries. He would not let them sink into despair. Every day he ordered the men to pack up and be ready to go in case the rescue ship came.

After the whalers rescued the small group on the other side of South Georgia Island they headed to Elephant Island, only to encounter pack ice which forced them to turn back. Shackleton organized another rescue attempt with a fishing vessel, but had to stop just 20 miles away from his crew. The men on Elephant Island were beginning to lose the struggle to survive. One man had his toes amputated due to gangrene, and another had a non-fatal heart attack. They were all malnourished. Time was running out.

On August 30, 1916 they finally saw the vision that they thought would never come—a ship! They quickly built a signal fire and were rescued by Shackleton himself. After 21 months in the desert known as Antarctica, they had all survived.

> It seemed to me that among his achievements, great as they were, his one failure was the most glorious. We had pierced the veneer of outside things. We had suffered, starved and triumphed, groveled down, yet grasped at glory, grown bigger in the bigness of the whole... (PBS, 2002).

Read the synopsis article from Outside Magazine. *Into Thin Air: A Personal Account of the Mount Everest Disaster* by Jon Krakauer OR Read entire article from Outside Magazine (see bibliography on p. 245). This account serves as a comparison with the Shackleton Expedition to show how working in a group can both succeed and fail.

INTO THIN AIR

> *Everest deals with trespassers harshly: the dead vanish beneath the snows. While the living struggle to explain what happened. And why. A survivor of the mountain's worst disaster examines the business of Mount Everest and the steep price of ambition (Krakauer, 1996).*[4]

Jon Krakauer reached the summit of Mount Everest on May 10th, 1996. Although he had been dreaming of this moment for quite some time, his five minutes at the top of the world were shrouded in a mental fog. After going 57 hours without sleep and almost no food, he was so cold and tired that he almost didn't care. That's what can happen at 29,000 feet above sea level.

There were many people trying to get to the summit that day. Krakauer had arrived earlier than most and was on his way down when he met a crowd at a bottleneck called the Hillary Step. At this point people must clip into a rope and ascend one-by-one. He had to step aside and let the 20 people going up pass by. It was a true traffic jam.

A few minutes earlier he had looked up at the peak he had just visited and noticed the weather was changing. It had been clear just an hour before; now clouds were beginning to move in. Jon would reflect on that moment a few days later after eight people had died and another had lost his right hand to gangrene on the mountain. How could this have happened? Although there were many amateurs on Everest that day, they were accompanied by veteran guides. Why did they keep going?

Among the dead were two guides, so it may never be known what really happened. The weather turned fast, and decisions made when one is over 25,000 feet can be clouded because of the thin air. Even though most people use oxygen tanks, human beings are not meant to live much above 5,500 meters (18,000 feet). Above 8000 meters is called the "dead zone," and it wreaks havoc on body and brain. Even at base camp, which is about 5300 meters (17,600 feet) above sea level Krakauer noted

> ...walking to the mess tent at mealtime left me wheezing for several minutes. If I sat up too quickly, my head reeled and vertigo set in. The deep, rasping cough I'd developed in Lobuje worsened day by day.... Cuts and scrapes refused to heal. My appetite vanished and my digestive system, which required abundant oxygen to metabolize food, failed to make use of much of what I forced myself to eat; instead my body began consuming itself for sustenance. My arms and legs gradually began to whither to stick-like proportions (1997, p. 68).[4]

As Krakauer waited at the Hillary Step that day, he noticed that most of the people ascending were part of a group led by Scott Fischer, an experienced guide from the Seattle area. He also noticed a couple of people from his own group, including Yasuko Namba. She would soon be the oldest woman to climb Everest, and the second Japanese woman to reach the highest points on each continent, also called the "Seven Summits." Reaching the Seven Summits is a goal of many mountaineers.

The last person in the traffic jam up the Hillary Steps was Fischer. Known for his energy and strength, he was a well-known and seasoned climber. He had even climbed Everest without oxygen. On this day, however, he looked tired and worn out. Once all had passed, Krakuer continued his descent to the South Summit. When he got there, the weather at the top of the mountain looked threatening. It was mid-afternoon, and soon things would take a turn for the worse....

Why were there so many non-climbers on Everest that day? It used to be that the world's tallest mountain was like the moon—virtually impossible to get to. Sir Edmund Hillary was the first Westerner to scale the peak. He was followed by Willi Unsoeld and Tom Hornbein of the United States. By the mid 1980's, Everest's easiest route had been climbed over 100 times, and in 1985 a wealthy Texan by the name of Dick Bass was guided up the mountain to become the first person to bag the Seven Summits. The attention he received caused many amateur climbers with enough money to hire guides to usher them to the top of the mountain. Once a seemingly impossible feat, it had become fashionable to climb Mount Everest.

The rest of the Seven Summits had also become popular. Businesses grew just to meet the demand. There was money to be made in guiding amateur climbers up the highest mountains in the world. On this expedition, New Zealand guide, Rob Hall, charged $65,000 per person to take people up the mountain. No one charged more, but no one was more respected for his skill and leadership. Jon Krakauer was one of Hall's clients on this trip.

On April 10th, they arrived at Everest Base Camp, which is located at 17,600 feet, or about 5,000 meters. In contrast to the rugged terrain, the camp was really a small city of tents comprised of 14 expeditions. They had most of the comforts of home, including a stereo system, phone, fax, and solar powered lights. They had hot showers and hired help who would even serve them hot tea while they stayed warm in their sleeping bags.

> [Rob] Hall's Adventure Consultants compound served as the seat of government for the entire Base Camp, because nobody on the mountain commanded more respect than Hall. Whenever there was a problem—a labor dispute with the Sherpas, a medical emergency, a critical decision about climbing strategy—people trudged over to our mess tent to seek Hall's advice. And he generously dispensed his accumulated wisdom to the very rivals who were competing with him for clients, most notably Scott Fischer (Krakauer, 1997, p. 60).

Both accomplished guides, Fischer and Hall, had very different styles. Fischer was more impulsive and encouraged his clients to go at their own pace. Hall insisted that his group climb together at all times, and had a system for helping climbers acclimate to the altitude. Styles notwithstanding, both guides had a business to run and were under pressure. This was Fischer's first guiding trip on Everest and he had something to prove. Hall had not gotten anyone to the top the previous year and also had something to prove so that his business did not wane.

Guided trips rely on Sherpas, native people who live in the Himalayas. They do everything from setting up tents and carrying equipment to setting up ropes and blazing the way for the less skilled climbers. Safety is always an issue, and the Sherpas make the ascent safer for the clients. There was another expedition of Taiwanese climbers who were on their own, lacking equipment and technical expertise to climb the mountain. On an earlier trip to McKinley in Alaska when they were preparing for the Everest climb, almost half of their group got caught in a storm and had to be rescued. One of the climbers died, while two had severe frostbite.

> The presence of the Taiwanese on Everest was a matter of grave concern to most of the other expeditions on the mountain. There was a very real fear that the Taiwanese would suffer a calamity that would compel other expeditions to come to their aid, risking further lives, to say nothing of jeopardizing the opportunity for other climbers to reach the summit. But the Taiwanese were by no means the only group that seemed egregiously unqualified. Camped beside us at Base Camp was a twenty-five-year-old Norwegian climber named Petter Neby, who announced his intention to make a solo ascent of the Southwest Face, one of the peak's most dangerous and technically demanding routes—despite the fact that his Himalayan experience was limited to two ascents of neighboring Island Peak, a 20,270-foot bump... (Krakauer, 1997, p. 94).

There was also a South African expedition that was led by Ian Woodall, who had lied about his experience. When this came to light, the experienced climbers in his group deserted, leaving only the members with little to no climbing experience.

> The solo Norwegian, the Taiwanese, and especially the South Africans were frequent topics of discussion in Hall's mess tent. "With so many incompetent people on the mountain," Rob said with a frown one evening in late April, "I think it's pretty unlikely that we'll get through this season without something bad happening up high" (Krakauer, 1997, p. 100).

Hall had chosen May 10th for the final ascent to the summit because of the weather patterns. The weather had a better chance of being good around this time of year, providing a window in which climbers could have a relatively clear day to make it to the top of the world. Because this was no secret there were quite a few groups wanting their chance at the summit. To coordinate things, they held a meeting to decide which groups would go and when. Hall's group, including Krakauer, would share May 10th with Scott Fischer's group. Most of the other teams promised to stay away from the top, with the exception of the team from South Africa. Woodall told the others that they would go whenever they pleased, and probably on the 10th.

Climbing a mountain of this caliber is full of risks. During the six weeks that Krakauer was on Everest, there had been a series of mishaps, including a Sherpa who fell and injured his leg, a climber who suffered a serious heart attack, another Sherpa who fell ill, and a climber who was hit with a large block of falling ice. It takes good systems, good judgment, and even some luck to get through a world-class climb unscathed. Experienced climbers, though, have rules under which they climb that help improve the safety. One of these rules is the "turn around time." This is the time that, no matter where a climber is, he or she turns back. A Swedish climber on the mountain that May decided to turn around at 2:00, even though he was within striking distance of the summit, believing that he would not have the energy to get back down if he continued. This was a display of good judgment. Hall's turn-around time for his group was 1:00 on May 10th.

The Hall and Fischer teams left for the summit on May 8th. Shortly thereafter a climber from another expedition took a fall. Although bruised and battered, he survived. His team left him in his tent to recover while they pressed on. Later, as he was descending, he fell over and died. This death caused people to pause, but since they were heading to the summit—their goal—they figured they could think about it later. They made it to the South Col, which was where they would prepare to reach the summit. They could get some needed rest in their tents before their attempt at 11:00 p.m. Most of the climbers did not sleep, though, due to the anticipation and the 50 mile-per-hour winds whipping their tents.

Krakauer contemplated his situation:

> There were more than 50 people camped on the Col that night, huddled in shelters pitched side by side, yet an odd feeling of isolation hung in the air.... In this godforsaken place, I felt disconnected from the climbers around me—emotionally, spiritually, physically—to a degree I hadn't experienced on any previous expedition. We were a team in name only, I'd sadly come to realize. Although in a few hours we would leave camp in a group, we would ascend as individuals, linked to one another by neither rope nor any deep sense of loyalty. Each client was in it for himself or herself, pretty much. And I was no different... (1997, p. 163).

On May 10th, at about 7:00 p.m., the weather cleared, and the expeditions were on their way to the summit by 11:35. Krakauer climbed quicker than most of the group, and ended up waiting for them to catch up because of Hall's rule that people stay together. He realized that being a client is different than working on one's own. The client loses the opportunity to be self-reliant and make decisions—the guide makes the decisions for the safety of all.

> Passivity on the part of the clients had thus been encouraged throughout our expedition. Sherpas put in the route, set up the camps, did the cooking, hauled the loads. This conserved our energy and vastly increased our chances of getting up Everest, but I found it hugely unsatisfying. I felt at times as if I wasn't really climbing the mountain—that surrogates were doing it for me. Although I had willingly accepted this role in order to climb Everest with Hall, I never got used to it. So I was happy... when, at 7:10 A.M., he arrived atop the Balcony and gave me the OK to continue climbing (Krakauer, 1997, p. 168).

Fischer's group was also climbing that day and Krakauer began to pass some of them. He noticed that one of the Sherpa's was "short-roping," or pulling, one of the clients up the mountain. It turns out that this client had only Everest left to complete her Seven Summits trophy, and had failed on two previous attempts. It is not known what actually happened, but it is possible that the Sherpa was either ordered by Fischer or bribed by the client to pull her up the mountain.

The turn around time is important for a variety of reasons, one of which is that oxygen is limited. Each climber had three oxygen tanks that would last a total of about 18 hours. To speed things up, the guides had decided to send up Sherpas at one of the bottlenecks to set up lines. This would speed up the process because people wouldn't have to wait for the equipment to be set. The Sherpa who was short-roping the client was one of the people who was to go ahead, which meant that the one left to do the work could not do it alone. Although Krakauer could have helped, he was forbidden to do so by his guide, Hall.

Because the ropes were not set, the climbers came to a stop at 28,000 feet while the ropes were being set. People got impatient. Some tried to push ahead, while others decided to turn back because it was becoming clear that they would not make it to the top before their turn around time. Given that they had spent tens of thousands of dollars and suffered from the severe conditions for many weeks, it must have been a difficult decision. But it turned out to be the right one.

There was another bottleneck at the South Summit, and the ropes were not set there, either. It turned out that the Sherpas who were there refused to place the ropes, ostensibly because they were tired of doing all the work for Fischer's group, of which they were not a part. Krakauer joined some other climbers to get the ropes up, but they had wasted a precious hour waiting.

Krakauer made the summit, and as stated earlier, it was less of a rush than expected. He stayed for five minutes and then noticed that his oxygen was dangerously low. It lasted until he had to wait at Hillary's Step. At this point, he saw his guide, Ron Hall, and thanked him. Hall was disappointed that most of his clients had turned back earlier, and that Fischer's clients were still moving forward. Krakauer started to panic at having no oxygen. Finally, another climber who had climbed without oxygen before, gave Krakauer his oxygen. Meanwhile it was clear that another guide, Andy Harris, was behaving irrationally, but it did not register with Krakauer that a guide might need his help. After all, the guides were there to help and protect him. "Given what unfolded over the next three hours, my failure to see that Harris was in serious trouble was a lapse that's likely to haunt me for the rest of my life" (1997, p. 188).

By 5:30 p.m. Krakauer was about 200 feet above his tent. He had one more obstacle in his way—a steep knot of ice that he would have to descend without rope—in a full-scale blizzard with the winds pounding at 70 miles per hour. Worried that in his deteriorated state he would make a critical error, he sat down to take stock. This is when Harris reached him, in even worse shape than before. Harris slid down the ice on his butt, and went the 200 feet flying down on his back. Krakauer could see him down below and thought he had at least broken something. A moment later, Harris was up and waving and stumbling toward the tents, which were intermittently visible in the blizzard. The last time Krakauer saw Harris, he was about 30 yards from the tents.

> I lunged into my tent with my crampons still on, zipped the door tight, and sprawled across the frost-covered floor too tired to even sit upright. For the first time I had a sense of how wasted I really was: I was more exhausted than I had ever been in my life. But I was safe. Andy was safe. The others would be coming into camp soon.... We'd climbed Everest. It had been a little sketchy there for a while, but in the end everything had turned out great (1997, p. 195).

But they had not all made it back, and many others were struggling for their lives:

- Pitman, the woman who had earlier been short-roped up the mountain, collapsed from altitude sickness. One of Fischer's guides dragged her down the mountain with a short-rope until she was able to recover and walk on her own.

- Mike Groom, another guide was descending with Namba and was short-roping another, Weathers, who was overcome with blindness. When Namba's oxygen ran out she became disoriented and refused to continue. Another guide, then, started dragging her down the mountain. This small group ended up wandering around in the blizzard in the dark and in wind chills of over 70 below zero. They were joined by two Sherpas who were also lost; no one had oxygen. They continued to wander aimlessly, hoping to stumble across camp. Finally around 10:00 p.m., they got everyone huddled together to wait for it to clear. Around midnight, it cleared enough for those who could walk to find their way back. They left behind four people who did not have the wherewithal to walk (Fox, Pitman, Weathers and Namba) along with one person (Madsen) to care for them. They waited for help.

  When Groom and company reached camp, they were greeted by Boukreev, a guide who had hurried back to camp on his own. There are questions about why he left his clients to fend for themselves, but he single-handedly rescued Fox, Pittman and Madsen. Later, a rescue party went for Weathers and Namba. Both, amazingly were barely alive, even though covered with ice. Their bodies were frostbitten or frozen. The decision was made to let them be and allow nature to take its course. Bringing them down would endanger more people, and they probably would not make it anyway.

Later that day Weathers stumbled into a camp, seemingly risen from the dead. "Initially I thought I was in a dream," he recalled. "Then I saw how badly frozen my right hand was, and that helped bring me around to reality. Finally I woke up enough to recognize that the cavalry wasn't coming so I better do something about it myself" (1997, p. 252). Still mostly blind, Weathers was able to make it back without tumbling thousands of feet to his death. Yes, sometimes it does take luck. Eventually, his right hand had to be amputated.

- Andy Harris, it turns out, never made it back to the tents, even though he was only 30 yards away. When Krakauer was alerted that Harris was not in his tent, he couldn't believe it. After retracing Harris' steps, he realized that instead of turning to go up a rise to the tents, he could have easily gone straight and down a 4,000-foot drop.

- Fischer, the energetic and fiercely independent guide, arrived at the summit at 3:30 p.m. He headed down with the Sherpa Lobsang. A short way down Fischer became disoriented, showing signs of hypothermia and cerebral edema (a build up of fluid in the brain). He started acting crazy, and Lobsang feared he would harm himself by jumping off the mountain. Lobsang short-roped Fischer until he could go no further. He then anchored Fischer to a ledge and was preparing to go for help. At this point, Gau, the leader of the Taiwanese expedition was being dragged down by three Sherpas. They tied the two disoriented men together and went for help. It was about 10:00 p.m.

  Four Sherpas went to rescue Fischer and Gau. Fischer was barely alive but unresponsive—they left him for dead. They revived Gau and descended with him.

- Guide Rob Hall and client Doug Hansen made it to the top after 3:00 p.m.—well after the turn-around time. Hansen had not made it to the summit on the last trip and was committed to making it this time. On the way down, Hansen ran out of oxygen and collapsed. At 4:30 p.m. Hall radioed for help and oxygen. Oxygen was nearby, but Harris, in his disoriented state falsely radioed in that the oxygen was gone. Instead, Hall tried to bring Hansen down without oxygen. It took them 12 hours to travel a stretch that is usually traversed in half an hour. At 4:43 a.m. Hall called in sounding weak and disoriented. Hansen was gone by this point, he probably fell to his death. At 9:30 a.m. two Sherpas ascended to try and rescue Hall. They made it to 28,000 feet, but had to turn back because of the weather. At 6:20 p.m., Hall was patched in a final time to his pregnant wife in New Zealand. This was the last communication anyone had with Hall. His body was found 12 days later.

Why had Rob Hall violated his own rule? Could it be that he had made a poor decision based on his business needs, rather than on his instincts as a mountaineer? Or did he let his emotions cloud his judgment because he wanted this client to make it this time?

Hall had been inordinately successful up until this point and had enjoyed good weather. Had he become cocky or had he run out of luck? Obviously judgment had a role to play in the disaster, and ignoring his own rule made all the difference. The apparent competition between Hall and Fischer may have, ultimately, played a role in both their and their clients' deaths.

> ... it is easy to lose sight of the fact that climbing mountains will never be a safe, predictable, rule-bound enterprise. This is an activity that idealizes risk-taking; the sport's most celebrated figures have always been those who stick their necks out the farthest and managed to get away with it. Climbers, as a species, are simply not distinguished by an excess of prudence. And that holds especially true for Everest climbers: when presented with a chance to reach the planet's highest summit, history shows, people are surprisingly quick to abandon good judgment. "Eventually," warns Tom Hornbein, thirty-three years after his ascent of the West Ridge, "what happened on Everest this season is certain to happen again" (Krakauer, 1997, p. 275).

Krakauer and the others are haunted by the events of May, 1996. How could things have been different, and what can we learn from their experiences?

Lesson 1.A Sequence: continued

Hold small group discussions about the two events.
In small groups, have students discuss what worked and did not work for both adventures. Try to come up with at least 5 ideas. Share ideas with the class. Consider how people worked together, how decisions were made, and what motivated the decision making.

Create a chart of qualities and attributes of productive and non-productive groups.
What behaviors and attitudes help to make groups productive, and what behaviors and attitudes cause them to be non-productive? For example:

Productive Groups	Non-Productive Groups
Cooperation	Competition
Shared responsibility	One person does all the work
Etc.	Etc.

Create a social contract for the class.

A. Ask each student to write in their journal a few things that are important to him/her about getting along in a group.
B. Have them share their ideas and put them together into a class social contract.
C. Discuss what each means to them. For example, if someone says "respect," ask her to share what she means by that.
D. Post the contract and briefly revisit it every day or so to assess how the class is doing with their social contract.
E. Add concepts as necessary.

Lesson 1.B Overview: Recognize Diversity Within the Group[5]

Focus:
Honor diversity
Analyze one's learning styles
Explore one's spectrum of multiple intelligences
Practice reflection

Step	Activity	Time	Label
1	Before doing any activities around diversity, revisit the social contract with the class.	5 - 10 min	
2	Explore the concept of emotional safety and confidentiality through the Hand Slap activity.	5 - 10 min	
3	Discuss/explore how being under attack inhibits everyone's ability to do their best.	5 - 10 min	
4	Complete the Multiple Intelligence Inventory.	20 - 30 min	self assessment
5	Facilitate the Human Bar Graph.	15 - 30 min	
6	Give each student a copy of the inventory and descriptions.	5 - 15 min	portfolio
7	Have a discussion about strengths and weaknesses.	10 - 30 min	
8	Facilitate the Categories and Line Ups activity.	10 - 20 min	
9	Facilitate the Cliques activity.	10 - 20 min	
10	Reflect on the idea of cliques using the processing questions from the Cliques Activity.	10 - 20 min	
11	Facilitate the Find Your People Activity.	10 - 15 min	
12	Have a discussion about diversity in the classroom and school.	10 - 15 min	
13	Hold an informative discussion about being an ally.	15 - 20 min	
14	Explore the concepts around being an ally through activities and reflection.	20 - 40 min	
15	Explore the concept of group identities through activities and reflection.	20 - 35 min	
16	Journal assignment.	10 - 20 min	assessment
17	Research Project. Research someone who has gained and lost because of his or her group identities.	varies	portfolio
18	Create and post a class chart of ways to honor diversity in the class and in our lives.	20 - 40 min	assessment

Time Frame: 3.5 - 6.5 hours, plus research time

Lesson 1.B Sequence: Recognize Diversity Within the Group

1.B sequence 1

Before doing any activities around diversity, revisit the social contract with the class. If the following are not incorporated into your ground rules, make a point to add them to the list:

Assume Good Intentions: There is opportunity for lots of discussion during class. There may be times when you disagree with something someone says, and may even be offended by it. It is very rare that someone is intentionally trying to be offensive. Assume that they mean well, and see it as an opportunity to enter into dialogue.

Ouch/Oops/Sorry: Sarcasm is commonly used as a way to show affection to people we know well. It, unfortunately, also has the side effect of being used as way to put people down. When groups are forming, or new people enter into a group, it is especially important to watch our use of sarcasm. If, for example, I say something sarcastic to you, you can say "ouch." What this does is inform me that I have said something hurtful (which I probably don't know). It then gives me the opportunity to say "oops," and make it right, "I'm sorry." Without this information I may never know that I said something hurtful, and cannot change my behavior. This information is a gift that can help me grow.

Right to Pass: You have the right to share what you are ready to share and do what you are ready to do. Anytime we are having a discussion, you have the right to not say anything at all and simply "pass." If we are doing an activity that may, for example, make a physical condition worse, please feel free to observe the activity, or do it in a way that will not cause you pain.[6]

1.B sequence 2

Explore the concept of emotional safety and confidentiality through the Hand Slap activity.

activity

Hand Slap

Focus: Appreciating diversity, perspective taking, empathy, safety
Materials: None
Time: 5 - 10 minuntes
Sequence: Deinhibitizer, Trust
Sources: Gary Hollander: www.garyhollander.com, Marilyn Levin: www.marilynlevin.com

Suggested Procedure

1. Explain to students that being under attack inhibits everyone's ability to do their best.
2. Each participant needs to find a partner.
3. Get into the position for playing the nasty hand slapping game so many of us played in grade school.

   For those of you who were spared this evil game, one person has their palms facing down and positioned directly over the other person's palms, which are facing up. The person in the palms up position tries to come around and slap the hand of the other person before they can pull their hand away.

4. If you are the slapper (palms up person) in this version of the game you also give the slapee (palms down person) four or five simple addition and subtraction questions while trying to slap their hands before they pull away.
5. Do this and then trade roles so each person gets to be the slapper and the slapee.
6. Once both people have experienced the challenge of attempting math while under attack, discuss the results.

Sample Processing Questions

- How many of you were able to successfully do math while under attack?
- What caused you to succeed or fail at the math?
- Did you try to help your partner or make it harder for them to do the math problems?
- What did you do to make is easier or harder for them?
- In what ways do you feel under attack from the external world? From your internal world?
- Are there things that you could do better if you didn't feel under attack or pressured by external or internal forces?
- Do you create an environment that makes others feel under attack? If so, how?
- Are there steps you would consider taking to change any of this? If so, what?

Facilitation Notes

Emphasize to students that this activity is not about hurting people, but about helping them understand the need to be safe. Watch for students that are being overly forceful with their slapping and ask them to be gentler.

Explain that this activity is a powerful reminder that none of us will ever do our best work while under attack. We never have to slap the person for them to know that we are a threat. The game is the set up. This is another piece of our cultural conditioning that interferes with our magnificence—persecution of others and of ourselves. In a world that emphasizes competition, judgment, criticism, and fear of difference, we have created an environment that is not conducive to the best work of anyone involved. Other themes to explore include:

Feeling Good About Ourselves: It makes sense to practice feeling really good about who we are on our way to believing it. When we come from a place that is deeply confident about our goodness as human beings, we have the strength to face up to the ways we are not our best selves without self persecution or defensiveness. Feeling truly good about who are we reduces our participation in self-pity parties, efforts to prove how good we are, and obsessions about what others will think of us. This saves an amazing amount of wasted time and energy that can be redirect toward making a difference in the world!

Feeling good about yourself is the best place from which to grow. Without this, it is too easy to shut down to the inhumanity that exists in the world, to be defensive about what feels like an inability to make real change in the world, and to be sucked into judgment instead of deepening the massive compassion needed to solve current world problems.

Self Attacks: The environment we spend the most time in is our internal environment (inside our own heads). So if we are doing things that are self-attacking, we are inhibiting ourselves from doing our best work. It is extremely useful to learn to end negative, critical, demeaning self-talk and replace it with self-talk that is supportive and empowering. When we are at peace with ourselves, we have no need to judge, criticize, or demean others.

1.B
sequence
3

Discuss/explore how being under attack inhibits everyone's ability to do his or her best.

Using the processing questions and facilitation notes from the activity, hold a discussion about being safe with each other. Talk about confidentiality—being safe does not end when they leave the classroom. Add ideas about confidentiality and safety to the social contract.

1.B
sequence

Facilitate the Multiple Intelligence Inventory (What are my learning strengths?).

A. Ask students to take out a piece of paper and a pen or pencil.
B. Have them write a "1" on their paper. (Caution: Don't identify the categories until done with survey.)
C. Explain that you are going to read a series of statements. Each time a statement is true for them, they should put a mark on their paper next to the #1. Later they will count them up.
A. Using the inventory list that follows, read all the attributes from the Verbal Linguistic area.
E. Now ask them to write a "2" on their paper.
F. Read the Logical Mathematical attributes and have them mark those that are true for them.
G. Continue with the rest of the Intelligences.

What Are My Learning Strengths? [7]

Research by Howard Gardner, from Harvard University, shows that all human beings have a "spectrum" of at least nine different types of intelligence, which he has named "Multiple Intelligences." Depending on your background and age, some intelligences are more developed than others. As each person learns and grows, he or she can develop the different types of intelligence. The important thing to remember is that everyone is smart in his or her own ways. This inventory is a way for you to focus on the ways you are smart.

Put a check mark beside each statement that is true about you.

1. **Word Smart (Verbal/Linguistic Intelligence)**

___ Like to write

___ Use words well when speaking

___ Enjoy telling stories and jokes

___ Have a good memory for names, places, dates, and other information

___ Enjoy reading

___ Like poems, puns, and tongue twisters

___ Like word activities like Scrabble®, anagrams, crossword, puzzles, and so on

___ Like to speak in front of groups

___ Find it easy to explain ideas to others

___ Often contact friends through notes and letters

2. **Thinking/Numbers Smart (Logical/Mathematical Intelligence)**

___ Think things out clearly

___ Can do math in head

___ Enjoy using computers

___ Like chess and checkers

___ Like to do experiments

___ Like working on thinking puzzles

___ Keep things neat and orderly

___ Like step-by-step directions

___ Like structure

___ Find it easy to solve problems

3. **Pictures/Images Smart (Visual/Spatial Intelligence)**

___ Like pictures and other visuals

___ See pictures in mind when thinking

___ Like mazes, jigsaw puzzles, and Lego® blocks

___ Enjoy drawing and designing things

___ Like maps and charts

___ Daydream a lot

___ Like creating art using different tools: chalk, paint, or markers

___ Like to rearrange a room

___ Like watching plays, musicals, and other performances

___ Can remember the way a room looks and feels

4. **Music Smart (Musical/Rhythmic Intelligence)**

___ Enjoy music a lot

___ Often sing, hum, or whistle songs to themselves

___ Play musical instrument or sing in a choir

___ Hear sounds others may miss—birds, crickets, bells in the distance

___ Like to have music playing all the time

___ Find it hard to concentrate while listening to the radio or TV

___ Have good rhythm to music

___ Like the rhythms of poetry

___ Like musicals better than dramatic plays

___ Find it easy to remember words of songs

5. **Body Smart (Bodily/Kinesthetic Intelligence)**

___ Cleverly mimic other people's movements and behaviors

___ Enjoy taking part in sports

___ Like to dance, act, do aerobics, martial arts, or mime

___ Move a lot when sitting on a chair

___ Like physical activities such as hiking, swimming, biking, or skating

___ Good with woodworking, sewing, or carving

___ Enjoy making things with hands

___ Like working with tools

___ Find it hard to sit still for long periods

___ Enjoy arts and crafts

This survey was developed by, and is used with permission of, Dell Coats-Erwin © 1999. The Multiple Intelligence name and concept was developed by Dr. Howard Gardner of Harvard University.

6. **People Smart (Interpersonal Intelligence)**

___ Am a leader in the neighborhood or school

___ Understand people very well

___ Have a lot of friends

___ Like to be with people

___ Try to solve disputes

___ Enjoy group games and/or group events

___ Care a lot about people and their feelings

___ Learn and perform best when working with others

___ Dislike working alone

___ Like belonging to clubs and other groups

7. **Self Smart (Intrapersonal Intelligence)**

___ Am deeply aware of inner feelings and thoughts

___ Have strong personality and will

___ Like to work on projects alone

___ Seem to live in own private, inner world

___ Have self-confidence

___ Act very different in style of dress and behavior

___ Put out a lot of effort when I believe in something

___ Like to be involved in causes that help others

___ Am very aware of what I believe

___ Believe that fairness is very important

8. **Nature Smart (Naturalist Intelligence)**

___ Care deeply about animals

___ When outside, closely notice sky, clouds, and plants

___ Enjoy growing plants

___ Like collecting rocks and seashells

___ Like going to the beach or walking in the woods

___ Like to watch fish in an aquarium for a long time

___ Care very much about the environment and endangered species

___ Believe it is very important to recycle

This survey was developed by, and is used with permission of, Dell Coats-Erwin © 1999. The Multiple Intelligence name and concept was developed by Dr. Howard Gardner of Harvard University.

___ Enjoy hiking and camping

___ Spend a lot of time outdoors

9. Deep Thoughts Smart (Existential Intelligence)

___ Enjoy looking at the stars at night and thinking about how everything fits together

___ Frequently think about the significance of my own life and my impact on others

___ Enjoy reading certain kinds of philosophy, religious, or spiritual literature

___ Like to figure out how things relate to each other

___ Sometimes have trouble focusing on details

___ Like to think about why things happen as they do

___ Like to read and learn about the meaning of life

___ Ask lots of questions that start with "Why?" and "How?"

___ Wonder a lot about why people die

___ Like to listen to sermons, discuss deep subjects, or both

_______ Verbal/Linguistic

_______ Logical/Mathematical

_______ Visual/Spatial

_______ Bodily/Kinesthetic

_______ Musical/Rhythmic

_______ Interpersonal

_______ Intrapersonal

_______ Naturalist

_______ Existential

1.B
sequence
5

Facilitate the Human Bar Graph.[8]

A. Set up a line on the floor using tape or rope. Assign one end as 0 and the other end as 10. Then choose one of the Intelligences, such as Verbal/Linguistic.

B. Have students line up according to how many of the attributes they have from the Multiple Intelligences Inventory they just did. If four people had 1, they would line up near the 0 one behind the other. Likewise, if three people had 5, they would line up one behind the other near the middle of the rope. It might look something like this:

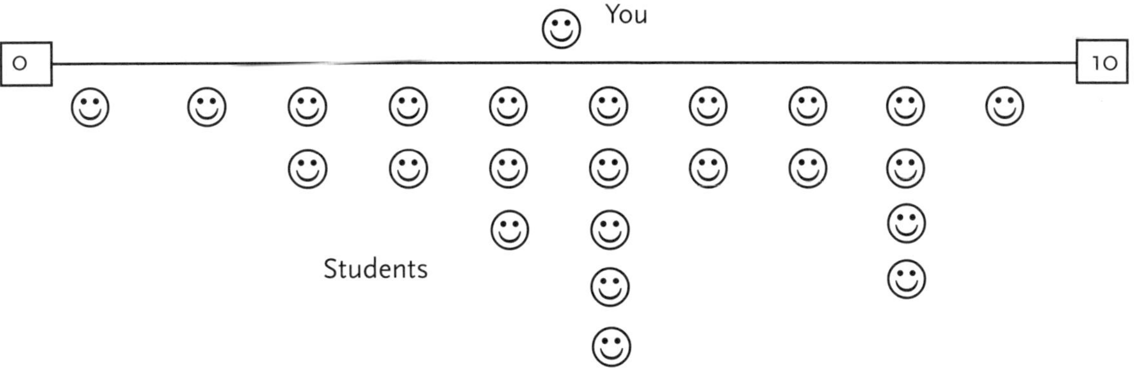

C. Read the description of the intelligence out loud.

D. Do the same with the other intelligences. Note how people move around.

Give each student a copy of the inventory and descriptions (following).

Have students highlight their strengths for their portfolios.

MULTIPLE INTELLIGENCES [9]

Linguistic: A person with linguistic intelligence is good at using spoken and written language, is good at learning languages, and relies on language to accomplish goals. Lawyers, speakers, writers, and poets are among the people with high linguistic intelligence. Linguistic learners:

- Like to: read, write and tell stories.
- Are good at: memorizing names, places, dates and trivia.
- Learn best by: saying, hearing and seeing words.

Logical/Mathematical: A person with logical/mathematical intelligence is good at analyzing problems logically, carrying out mathematical operations, and investigating issues scientifically. Mathematicians, logicians, and scientists exploit logical/mathematical intelligence. Logical/mathematical learners:

- Like to: do experiments, figure things out, work with numbers, ask questions and explore patterns and relationships.
- Are good at: math, reasoning, logic and problem solving.
- Learn best by: categorizing, classifying and working with abstract patterns/relationships (Gay, n.d.).

Musical: A person with musical intelligence is good at performing, creating, and appreciating musical patterns. Choreographers, singers, instrumentalists, composers possess musical intelligence. Musical learners:

- Like to: sing, hum tunes, listen to music, play an instrument and respond to music.
- Are good at: picking up sounds, remembering melodies, noticing pitches/rhythms and keeping time.
- Learn best by: rhythm, melody and music (Gay, n.d.).

Bodily/Kinesthetic: A person with bodily/kinesthetic intelligence is good at using the whole body or parts of the body (like the hand or the mouth) to solve problems or make things. Dancers, actors, and athletes have significant bodily/kinesthetic intelligence, along with crafts persons, surgeons, bench-top scientists, mechanics, and many other technically oriented professionals. Bodily/kinesthetic learners:

- Like to: move around, touch, talk and use body language.
- Are good at: physical activities (sports/dance/acting) and crafts.
- Learn best by: touching, moving, interacting with space and processing knowledge through bodily sensations (Gay, n.d.).

Spatial: A person with spatial intelligence is good at recognizing and manipulating patterns of wide space (such as those used by navigators and pilots) as well as patterns of smaller areas (such as those in which sculptors, surgeons, chess players, graphic artists, or architects work). Spatial learners:

- Like to: draw, build, design and create things, daydream, look at pictures/slides, watch movies and play with machines.
- Are good at: imagining things, sensing changes, mazes/puzzles and reading maps, charts.
- Learn best by: visualizing, dreaming, using the mind's eye and working with colors/pictures (Gay, n.d.).

Interpersonal: A person with interpersonal intelligence is good at understanding the intentions, motivations, and desires of other people and is therefore good at working well with others. Salespeople, teachers, clinicians, religious leaders, political leaders, and actors all need acute interpersonal intelligence. Interpersonal learners:

- Like to: have lots of friends, talk to people and join groups.
- Are good at: understanding people, leading others, organizing, communicating, manipulating and mediating conflicts.
- Learn best by: sharing, comparing, relating, cooperating and interviewing.

Intrapersonal: A person with intrapersonal intelligence is good at understanding the self and has a good working model of the self—including one's own desires, fears, and capacities—and can use such information effectively in regulating one's own life. Intrapersonal learners:

- Like to: work alone and pursue own interests.
- Are good at: understanding self, focusing inward on feelings/dreams, following instincts, pursuing interests/goals and being original.
- Learn best by: working alone, individualized projects, self-paced instruction and having own space.

Naturalist: A person with naturalist intelligence is good at recognizing and classifying numerous plant and animal species, is comfortable in nature, and may be good at caring for, taming, or interacting subtly with various living creatures. Environmentalists, hunters, fishermen, farmers, gardeners and cooks utilize naturalist intelligence. Naturalist learners:

- Like to: be outside with animals, geography, and weather; interacting with the surroundings.
- Are good at: categorizing, organizing a living area, planning a trip, preservation, and conservation.
- Learn best by: studying natural phenomenon, in a natural setting, discovering how things work (Gay, n.d.).

Existential: A person with existential intelligence is good at locating the self in the context of the farthest reaches of the cosmos—the infinite and infinitesimal—and is comfortable pondering such issues as the significance of life, the meaning of death, the ultimate fate of the universe, and other profound experiences such as love and artistic creation. Existential learners:

- Like to: ponder, have time to think, and be in places that are quiet.
- Are good at: seeing the big picture, identifying with feelings, understanding how things fit together, networking.
- Learn best by: having time to reflect, talking things through with others, knowing the context of information.

Have a discussion about strengths and weaknesses.

Ask students about how their strengths and weaknesses have been helpful or difficult for them. Here are some sample questions:

- Do you sometimes feel alone, without allies?
- Do you sometimes feel like you are able to use their strengths, or not use them?
- Do they see strengths in the group or in other group members? What does that tell them?
- Do you want others to appreciate/respect your intelligences? How do you feel when they don't?
- How good are you at respecting/appreciating other people's intelligences when they differ from yours?
- What types of people are easy for you to appreciate? What types are not so easy?
- What makes it easy or difficult for you to appreciate the intelligences or styles of others?[10]

1.B sequence

8

Facilitate the Categories and Line Ups Activity.

activity

Categories and Line Ups

Focus: Mixing with others, learning more about each other, appreciating diversity, perspective taking
Materials: None
Time: 10 - 20 minuntes
Sequence: Ice Breaker
Sources: See "Line up Like This – No Line up Like That" in *Adventure Education for the Classroom Community* by Frank and Panico, "Find Your Place" and "The Line Forms Here" in *Adventures in Peacemaking* by Kreidler & Furlong, *Journey Toward the Caring Classroom* by Frank, "Chronological Line Up" in *Silver Bullets* by Rohnke, "Where in the Circle Am I" and "Name by Name" in *Quicksilver* by Butler and Rohnke, and "Line Ups" in *Teamwork and Teamplay* by Cain and Jolliff, Marilyn Levin.

Suggested Procedure

1. Clear away an area so that people can move around with ease.
2. LINE UPS: Ask the class to line up according to:
   - Alphabetical order by first name
   - Alphabetical order by last name
   - Alphabetical order by mother's first name
   - Birthday
   - Shoe size
   - Height (shortest to tallest)
   - Hair color (darkest to lightest)
   - Skin color (darkest to lightest)
   - Thumb size (shortest to longest), etc.
3. CATEGORIES: Have them get into groups according to a certain category:
   - Number of siblings (count step and half bothers/sisters)
   - Season in which they were born
   - Favorite ice cream
   - Favorite day of the week
   - Eye Color

- Types of pets at home
- Plans for after high school
- Shirt color
- Types of holidays you might celebrate in winter? (Hanukkah, Christmas, Kwaanza, Ramadan, none...)
- What generation are you in this country?

4. Each time take a moment to notice the diversity in the room. Make and take comments about what people observe. Maybe there are a lot of chocolate ice-cream lovers in the room, or everyone seems to like the same day of the week. Why is that?
5. Transition to looking at groups and how people become identified (either by oneself or others) with those groups:
   - A group you arc proud to be a member of
   - A group you may not be proud to be in
   - A group you get praised for being in
   - A group you get teased for being a member of
   - A group you used to be in but are not in any longer
   - A group you are not in now but expect to be in at some point

Sample Processing Questions

- What do we seem to have in common in this class?
- What are some of our differences?
- How do you like the idea of this kind of diversity?
- What are some of the other things that make you unique?

Facilitation Notes

This activity can be repeated throughout the year. Each time, the categories and line-ups can be used to explore diversity issues to a deeper level. Have students create their own categories and line ups. Then use one as a warm up for class.

Sometimes a student may not know the answer to a question, such as mother's first name, or what generation she or he is in this country. Make sure that students know that they have the right to pass, and that they can choose to observe, use another family member's name, or join any group they wish. The object is not to be legally correct about such things, but to learn more about others.

9

Facilitate the Cliques Activity.

This activity helps students explore their understanding of, and participation in, cliques.

activity

Cliques

Focus: Perspective taking, empathy, stereotyping
Materials: None
Time: 10 - 20 minuntes
Sequence: Trust
Sources: Marilyn Levin

Suggested Procedure

1. Have the group name different cliques that exist in their school. Usually, half a dozen to a dozen cliques get named. Examples include: "sportos, preps, dirtballs, stoners, and nerds."
2. Choose five or six clique categories by selecting the main ones or by combining some of the cliques that are similar.
3. Ask the group to line up in horseshoe style according to how much they identify with a particular clique—say sporto. An example may sound something like: *"I want you all to line up next to each other, shoulder to shoulder, in a horseshoe shape based on how much you see yourself as a sporto. The most completely, totally sporto person lines up on the right side to the most completely not-sporto person on the left side with everyone in between lining up where they see fit."*
4. Once people are lined up, check in with the different parts of the line up. Those who classified themselves as strongly feeling part of the clique, those who had "middle of the road" or ambivalent feelings, and those who strongly considered themselves not associated with the clique.
5. If a group is small enough and you have time, consider checking in with each person. The questions to ask participants include: *"So, why did you end up here?"* This causes people to explore how they identified as that label or not, and what that label means to them. Then, you can start exploring the ways in which people agree on the label as well as the different meanings the label has for different people.
6. After completing one "line up" and discussion, then move on to the next clique label.

Sample Processing Questions

- What did you notice about the different line-ups? How did people sort themselves out?
- If most of us relate in some way to all of the cliques then why do we feel the need to segregate?
- Since we all have some connection to most other cliques, what makes us tease and target each other for being different?
- What would you gain if you decided not to play along with the clique systems at school?
- So what might you want to do differently now around the cliques in your school?

Facilitation Notes

One of the ways that "isms" (racism, sexism, etc.) are perpetuated is when people ignore teasing and put downs when they occur. It is important to be alert for any teasing or put downs (they can sometimes be quite subtle) during this activity. Remind students about the social contract and the idea of safety before starting. You and students can create a quick agreement to acknowledge any put downs or teasing right away. This means that everyone in the group is responsible for keeping the environment safe. Depending on the situation, you can stop to acknowledge it and move on, or take a moment to talk about what happened.

An organization called **Playboard** in Northern Ireland offers some advice when dealing with sectarianism. Their strategies are useful for dealing with all types of "isms." Among other ideas, they stress the following:

- Do not allow any discriminatory practice, language, or behavior to go unchallenged. This should cover race, culture, disability, color, sexual orientation, religion, gender, age, class, etc.
- Give students consistent messages—always deal with sectarian behavior.
- When stopping sectarian behavior explain why it is unacceptable. Further exploration of the incident may also be necessary.
- Be willing to let students discuss issues among themselves—they can learn a lot from each other and from the process itself.

1.B sequence 10

Reflect on the idea of cliques using the processing questions from the Clique Activity.
Before asking the questions, it may be helpful to define clique. First, ask the students how they define the term. Cliques, by definition are exclusionary. They may have many of the characteristics of a community—where people come together around common goals and interests, but outsiders do not feel welcome. Talk about how this idea of excluding others can be an obstacle for creating a feeling of community with a variety of people.

Another side to cliques is the idea of stereotyping. Just because someone is a member of a particular group, does not mean they represent that group. Stereotyping keeps people from seeing the person as an individual—one who possesses a variety of intelligences, talents, beliefs, emotions, etc.

1.B sequence 11

Facilitate the Find Your People Activity.

activity

Find Your People

Focus: Working together, mixing, friendship
Materials: 3-dimensional shapes of different colors. You can make your own out of wood or go to a learning shop and buy a container of those large plastic buttons that come in geometric shapes.
Time: 10 - 15 minuntes
Sequence: Ice Breaker, Deinhibitizer
Sources: I learned this from Jim Cain at a conference in 2004.

Suggested Procedure

1. Everyone takes a shape without looking at it or showing it to anyone else. They must also keep the piece to themselves (e.g., not pass it off to anyone else).
2. They are then instructed to "find your people," by only talking and describing their piece.

3. When everyone is in a group, look to see what shapes they have.
4. After they can see the shape (and, thus, have more information) ask students to find another way of "finding your people." They might then regroup according the color of the piece they have.

Sample Processing Questions

- How did you know that these were "your people?"
- Now that you can see the pieces, how might we regroup?
- What do all of these pieces have in common?
- What do we all have in common?
- How can we find out the things we might have in common with others?

Facilitation Notes

This activity can be used to remind us that there is more to people than what we see. Challenge students to be open to the fact that there are people out there who are friends waiting to happen. It takes some effort and risk, sometimes, to get past the obvious and seek out what they might have in common with others.

1.B sequence 12

Have a discussion about diversity in the classroom and school.

Using the foundation of the previous activities and reflection, talk about the following:

- What kinds of diversity are there in this classroom? In this school?
- What groups are there in the school? How can you tell who belongs to what group?
- How do they show their group identity?

Hold an informative discussion about being an ally.

When someone does not belong to a particular identity group, it is possible to be an ally. A definition of "ally" is: "A person who collaborates or associates with others." (Chappelle & Bigman, 1998) Marilyn Levin, a consultant in the area of diversity education describes the totality of being an ally:

> We generally think of an ally as a person who stands up for and supports others who are being picked on or put down in some way. Taking action to interrupt mistreatment is something that almost always makes sense. In doing so, you demonstrate your commitment to making the world a better place.
>
> While thinking of an ally as someone who helps those who are mistreated is helpful, it is even more productive to think of an ally as someone committed to a world that works for everyone. This way as an ally you stand up for everyone, you interrupt hurtful behavior toward anyone, and you become an ally for the causes that benefit us all.
>
> So, be an ally for both sides—for the bullied and the bully. Be an ally for justice, for fairness, for non-violence, and for a sustainable, peaceful world for all of us. Interrupt hurtful statements or demeaning behavior toward anyone. Interfere when anyone tries to make anyone else their enemy.
>
> It can seem odd to say that we should be an ally for the bully, for those who cause harm, for the inconsiderate person who lashes out at others. The reason this makes sense is because no one is born to hate and to mistreat others. People become capable of mistreating others because they have been hurt themselves and because our culture teaches us to judge and

criticize. People have to lose touch with their humanity and compassion to be capable of being hateful and hurtful. To be capable of mistreating others is very damaging to us as human beings because deep down we want to have cooperative, caring relationships with others. If we love ourselves fully we do not need to be hateful or hurtful to anyone else. So when we interrupt mistreatment we do so for the bullied person and for the bully because it is helpful and healing to do so for both parties. We do so for ourselves because of our commitment to a world that works for everyone.[11]

In their book, *Diversity in Action* (1998), Chappelle and Bigman offer their thoughts on being an effective ally:

- Within and between groups, we are all different. That should be both understood and celebrated.
- We all are capable of being effective allies for people from groups other than our own.
- You are critical in fighting for your own and other people's freedom from oppression.
- Apparent rejection of your offer(s) to be an ally to another group frequently stems from larger social issues, and not your specific offer.
- Allow for and be patient about differences in communication styles between your own and other groups.
- Understand that people are the experts on their own life experiences. This doesn't mean you can't understand their experiences or use your own to find solutions, but they know their own environment and experiences better than you.
- Whenever possible, actively ask and seek understanding of oppression experienced by members of other groups.
- Appreciate and, when appropriate, help people take pride in the history of their own group.
- In your quest for better understanding, allow for and learn from disagreements. Do not allow this process to prevent or inhibit you from being an effective ally. Remember, your intent and the outcome you receive don't always match.
- Language and behaviors that are comfortable for you are often cultural. Understand that what may be comfortable for you may make people from different groups uncomfortable (Chappelle & Bigman, 1998).[12]

activity

Perspective

Focus: Perspective taking
Materials: None
Time: 5 - 10 minuntes
Sequence: Ice Breaker
Sources: Marilyn Levin; Laurie Frank first learned this activity from Gloree Rohnke in 2002.

Suggested Procedure

1. Introduce the activity: "Let's consider some of the cultural conditioning that interferes with our ability to come at life and respond to others from the very best of who we are."
2. Have students hold one hand as high in the air as possible with their index finger pointed at the ceiling.
3. Draw a clockwise circle with this finger in the air.
4. Continue drawing this clockwise circle in the air with their finger pointed toward the ceiling and slowly bring their hand down to the point where they are circling at the level of their stomach.
5. Look down and notice that the circle is now going counter clockwise.
6. So what happened? Perspective shifted, the circle's direction did not.
7. Explain that the only thing that changed is that their perspective shifted. They went from looking up at the circle to looking down at it. This gave them opposite views on what was occurring. This is a reminder that we can accept other views or perspectives as accurate AND our views and perspectives as accurate at the same time, even if they seem to be in contradiction.

Sample Processing Questions

- How open to others' perspectives do you think you are?
- How open do you think others are to your perspectives?
- How hard would it be for you to change your perspective on some of your most deeply held beliefs or positions? What would it take to change your mind?
- What are some things that we used to believe a long time ago that we now know are not true?
- What might we believe now that later on we will discover is not accurate?
- When being an ally, why is it important to be open to different perspectives?

Facilitation Notes

This activity demonstrates the power of perspective. This activity is ideal to use as an introduction to other activities where you want to encourage people to be open to exploring other's perspectives, beliefs and ideas.

Deepak Chopra explains that there is no such thing as reality that is independent of who perceives the reality. Take a flower for instance. Human beings (who have sight) visually perceive the color and texture of the flower. A bee will not see the flower in the same way humans do. It will experience the flower as ultraviolet wavelengths. A bat will have yet another perception of the reality of a flower as the echo of ultrasound. So whose reality is accurate—well all of these and more are completely accurate depending on who is doing the perceiving of the reality.

So we constrain ourselves when we decide our perspectives are the right ones, when we limit our access to a broad range of wisdom and when we fail to embrace multiple ways of knowing and being. Certainly, if history is any indication, parts of what we now KNOW to be true or fact will turn out to be incomplete, inaccurate or just plain wrong. So it can be useful to loosen our grip on what we KNOW to be true and be open to wisdom that differs from ours. We get to remember that we are most often not aware of the paradigms we function within—it is a lot like trying to explain the reality of a world that is orange without noticing that we have on orange glasses.

It will be essential to use the wisdom of all cultures (including the ones that seem wrong to us) and to respect our inter-connectedness to solve the problems of our time in this global community. This will require that we learn to create space for embracing what will seem to be mutually exclusive beliefs and practices. Learning to hold multiple, seemingly conflicting truths is an essential skill to coexist in a global world.[13]

activity

Mind Power

Focus: Empowerment, taking action, integrity
Materials: One-quarter inch flat metal washer and a fifteen-inch string for each person. Tie one end of the string through the hole in the washer.
Time: 5 - 10 minuntes
Sequence: Deinhibitizer
Sources: See *Activities That Teach* by Jackson

Suggested Procedure

1. Everyone should have a washer that is tied to a string.
2. Hold the string out in front of you about level with your forehead.
3. Still the washer at the end of the string with your free hand.
4. By using your focus and concentration look at the washer and see of you can get it to move side to side while keeping your hand as still as humanly possible.
5. Once you've succeeded at this, try making the washer move front to back.
6. Then when you are ready for the next step try to get the washer to move clockwise or counter clockwise in a circle.

Sample Processing Questions

- What happened when you concentrated on making the washer move?
- What do you think made the washer move? Where does that power come from?
- What are the powers we need when being an ally? Where does that power come from?
- What makes it challenging to be an ally?
- What are ways to be a good ally?

Facilitation Notes

Being an ally means that we may end up in a situation that is difficult, challenging, and scary. It means having the integrity to stand with others because we believe it is the right thing to do. This activity is our reminder that we have amazing hidden capacities. We have many talents, skills, and abilities of which we may not be aware. The ability to take courageous action that is in line with what is right in our hearts and minds is a capacity we all possess.[14]

Pairs Tag

Focus: Honesty, rules, failure
Materials: Boundary markers
Time: 10 - 20 minuntes
Sequence: Deinhibitizer
Sources: See *Diversity in Action* by Chappell & Bigman, and "Pairs Squared" in *Quicksilver* by Rohnke & Butler.

Suggested Procedure

1. This activity needs enough space to move around—best in a gym or outside. A classroom works if people are careful to walk slowly.
2. Have people get into pairs and explain that each pair should pick one person to start as "it."
3. Explain that they will play tag with only the other person in the pair.
4. Explain that, once tagged, participants must put their hand on their head and turn around twice before they can go after their partner. Let them know that if this will make them throw up they should not do it and do something else in its place, like two jumping jacks.
5. Explain the rules of the activity: No running. And you can only be in physical contact with the person in your pair and the floor (no other people or items in the environment).
6. Before starting the tag game say, "We'll play for a short time, and I'll end the game with a count down from ten to zero—at zero you do not want to be it."
7. Answer questions and let the play commence.
8. You want some of the rules to be broken. If your group is not breaking rules by the time you count down, do so slowly and remind them they do not want to be it! This will get at least a few folks running and bumping into others.
9. Circle up and ask them to: "Raise your hand if you saw rules being broken." "Raise your hand if you broke a rule." Then ask, "What rules were broken?" and take comments from the group.
10. You can explain that the point of the activity was to set the participants up so that they would struggle to honor the rules. Now we can use what happened in the activity to explore topics like honesty, rule breaking, and setting people up to fail.

Sample Processing Questions

- How do we set others up to fail?
- How do we set ourselves up to fail?
- What are examples of ways we set others or ourselves up to fail?
- What can we do to stop doing this to others and ourselves?
- Are there ways that we set up our allies? What do these look like? How can we stop doing this?
- What can we do to support allies, rather than set them up?

Facilitation Notes

Being an ally is a choice. Sometimes people choose to set people up rather than be supportive. This can happen due to peer pressure, confusion, misunderstanding, or expectations within and between cliques. Reflecting on, and planning for, possible difficult situations can help people make conscious decisions when the need arises.[15]

Explore the concept of group identities through activities and reflection.
There are a plethora of identities to which people ascribe: race, ethnicity, religion, gender, sexual orientation, age, physical ability, mental ability, appearance, and class are ones that are most recognizable. Like a spectrum of intelligences, people also have a spectrum of identities, which contribute to the uniqueness of each individual.

activity

Exploring Our Diversity Ups and Downs

Focus: Perspective taking, differences and commonalities, exploring identity
Materials: A chair for each person
Time: 10 - 15 minuntes
Sequence: Deinhibitizer, Trust
Sources: Marilyn Levin; this activity was developed and used at the National Conference on Community and Justice Camp Any Town in Wisconsin starting in 2000.

Suggested Procedure

1. Start with people sitting in chairs, ideally arranged so everyone can see everyone else.
2. Introduce the activity with this description:
   We each represent diverse feelings, experiences and heritage. Diversity is not always about someone else. Some aspects of our identity are given. We can't change them. Others are within our control or influence. The identity we think of as our own is shaped in part by our experience while our experience is determined in part by the identities others attach to us. Too often we fail to recognize the diversity represented by any group or individual. Categorical assumptions and stereotypes lead us to perceive sameness in the midst of difference.
3. Inform participants that the purpose of the exercise is to explore our diversity and have fun. Remind participants that it will be important for everyone to listen to your directions. Ask everyone to remain silent during the exercise. (If your group wants to acknowledge people for standing, it can work well to use the silent "applause" style used by the deaf and hard of hearing community. Hands are up on either side of your face with palms facing out. You rotate your wrists so that your palms shift back and forth from facing out and facing in).
4. Inform participants that you will be announcing various social groupings. If they identify themselves as a member of an announced group they are invited to stand (or raise their hand if unable to stand) or witness others who identify with their group. Then participants take their seat again. Create other options if any students have mobility restrictions (e.g., raising hands or saying "that's me" instead of standing).
5. Secure commitments regarding confidentiality from participants or if confidentiality cannot be reasonably assured then let participants know they should make choices accordingly.
6. Tell the group that claiming membership in some groups can feel uncomfortable or unsafe. In fact real threat of physical or emotional harm can exist for individuals who openly claim all aspects of their identity. **Remind the class about the ground rule giving everyone a right to pass. The exercise extends an invitation to stand. No one is under an obligation to do so. Failing to claim an identity is in NO way an indication of shame about that identity.**

Sample Processing Questions

- What was that experience like for you?
- What thoughts and feelings did it bring up for you? Was anything surprising to you?

- What did you notice about the diversity (or lack of) in the group?
- Was it challenging in this activity to claim any of your identities?
- Was it empowering in this activity to claim an identity?
- Anyone have questions about an identity that was new to him or her?
- Anyone feel like they now wish they hadn't claimed an identity in this exercise?
- Anyone feel like you are stereotyped for your identities?
- How can you go about making this group a safe place for all of our identities?
- What will you need to refrain from doing so that others feel safe in this group?
- What can you do individually to make all people feel valued here?
- How can you get support in being open to people who are different from you?

Facilitation Notes

The categories to be announced in this exercise can be tailored to your group's needs. Begin with simple categories that participants will not perceive as threatening. Keep the exercise moving. Be aware of the attention and energy of your group. Before announcing any categories that may provoke anxiety in participants, state again the voluntary nature of this exercise.

Interrupt any discounting behavior—snickering, booing, inappropriate laughing. First, attempt to remind participants about the emotional safety goals of the activity and the commitments the participants have made to each other around emotional safety. If this does not work, a next step can be to own up as the facilitator that your decision about this being an appropriate time to do this activity was inaccurate. If you do this without blaming yourself or the group, it can actually get the group invested in continuing.

You will not want to use this activity until a degree of safety has been assessed so that participants will not be targeted during or after the activity for their disclosures.[16]

activity

Identity Line Ups

Focus: Perspective taking, differences and commonalities, exploring identity, communication
Materials: None
Time: 10 - 20 minuntes
Sequence: Ice Breaker, Deinhibitizer, Trust
Sources: Marilyn Levin

Suggested Procedure

1. Explain that the group will be asked to "line themselves up" based on certain characteristics or topics.
2. Participants should end up in a U or horseshoe shape instead of on actual line so they can easily see and communicate with each other. Participants will line up shoulder to shoulder with no grouping.
3. After giving instructions, the facilitator allows the group to place themselves on the horseshoe with as little or as much conversation as they wish, depending on the time constraints. In general most groups will not take much time to place themselves.
4. Begin by doing some regular line ups.
    - Eye color (light to dark)
    - Eating habits (healthy to unhealthy)

- Favorite animal (by size of animal)
- Any personality trait—outgoing, patient, risk taking—(least to most), etc.

5. You can also make stipulations for any of these like no talking, only use animal sounds, only use hand symbols to communicate, etc.
6. Once participants are in the line up, the facilitator proceeds to ask for the label or the explanation of the placing. Depending on group size and time constraints, this will be done by asking participants one by one or by asking participants in a certain part of the spectrum who would like to volunteer to comment. The processing is generally done in the horseshoe but if a lengthy discussion ensues, you can sit down and circle up if desired.
7. Continue line-ups with topics that will give students an opportunity to explore identity.

Sample Processing Questions

- How easy or difficult was it to find your place in the line?
- What did you notice about the diversity of identities in this group?
- Why might it be too risky to disclose a hidden identity?
- What can we do over time to help people feel free to share their full spectrum of identities? How do we build that level of trust?
- In what situations do you feel free to "be yourself?" How is that empowering?

Facilitation Notes

Line-ups is a wonderful tool—simple to use and amazing discussion openers.

Many of these versions ask that people disclose hidden identities. It is VITAL that group emotional safety be adequate to make such disclosures safe. Inform the participants before, and remind them during, the activities that they should do whatever they need to remain safe, even if that means lying in words or behavior to keep an identity hidden. And that choosing to keep an identity hidden does NOT mean that they are ashamed of that identity. It can just mean that the potential costs of disclosure are not worth the risks for them in this instance.

In addition, facilitators should not embark on the riskier versions of these activities until they know they have the knowledge and the skills to do so.

With line ups it makes sense to insist on participants standing shoulder to shoulder (not allowing bunching). This pushes the issue of people not always being able to stand where they see themselves fitting on the spectrum. They end up being forced into standing where they fit in comparison to others in the group instead of where they see themselves on the spectrum of the topic at hand.

Line Up Suggestions

Alphabetical Label Line Ups:

Pick any number of topics and have participants choose whatever label they think fits them within the topic area. Then have them line up in a horseshoe based on their label alphabetically. Examples: choose a label that fits you around school cliques, race/ ethnicity, age, religion/spirituality, ability, class/income, gender, stereotypes, etc.

Give people complete freedom to select whatever label comes to mind, pointing out if needed that they likely have multiple labels that would fit and can change their minds at any point in the activity. Then ask the group to tell others their label and have a great discussion about labels and identities.

You can also have participants pick a label that fits them that they like, or a label that fits them that they hate, or a label that they have that's confusing or conflicting to them, or a label that someone else has given them that they like or dislike, or does or doesn't fit in their mind.

Diversity Related Line Ups:
Pick from topics listed below and have participants line up within the topic area.

Not risky: age (youngest to oldest), comfortable with labels (most to least)

Can be, but may not be, risky: masculine to feminine, politically progressive to conservative, religious or spiritual (most to least)

Sometimes risky: Ability (most able to least able) Class (working class to owning class), Privilege (most to least).[17]

Journal assignment.

Choose one or more: How has your uniqueness been helpful or hurtful to you in school and in your life? How have you been an ally, and/or wish to be an ally with others? Who do you look down upon and how would your life be different if you decided to become their ally? What have you been picked on for and what would you like your allies to say or do about this?

Research Project. Research someone who has gained and lost because of his or her group identities.

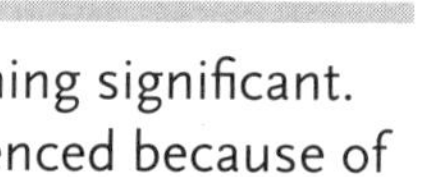

Have students do research on a person of interest who has accomplished something significant. Write a paper or give a presentation about the costs and benefits that they experienced because of their group identities, and how he or she dealt with the circumstances.

Suggestions are: Helen Keller, Malcolm X, Vincent Chin, Martin Luther King, Jr., Matthew Shepherd, Cesar Chavez, Gloria Steinem, Maya Angelou, Christine Craft, Tiger Woods, Mel Tillis, Fannie Lou Hamer, Maggie Kuhn, John Glenn, Martina Navratilova, Leonard Peltier.

Create and post a class chart of ways to honor diversity in the class and in our lives.

Marilyn Levin reminds us that "honoring diversity is really about a commitment to ourselves—not an obligation to others. It is about who we are as human beings and who we are willing to become. If we accept the challenge of becoming amazing allies we win because we get our best selves back—our most developed compassionate, loving, powerful selves."[18]

Discuss the following questions:

- What are ways to honor diversity in this class and in our lives?
- How can we be allies with each other?

Ask students to write down their ideas in their small groups and share with the class. Decide on the top 5 to 10 strategies and post them next to the social commitment.

Lesson 1.C Overview: Differentiate Between Cooperation and Collaboration

Focus: Build group consensus about the nature and functions of collaboration
Practice reflection

Step	Label	Activity	Time
1		Create base teams.	5 - 10 min
2		Facilitate the Human Treasure Hunt activity in base teams.	15 - 20 min
3		Facilitate the Puzzles problem-solving activity that requires cooperation.	10 - 20 min
4	assessment	Complete a large group "Pi Chart" for Cooperation.	15 - 20 min
5		Facilitate the ABC Pyramid activity to explore the notion of collaboration.	20 - 30 min
6		In base teams, answer the processing questions for the ABC Pyramid activity.	10 - 15 min
7		Share information about cooperation and collaboration.	15 - 20 min
8	assessment	In base teams, create definitions for "cooperation" and "collaboration."	15 - 30 min
9		Read some established definitions of cooperation and collaboration.	10 - 15 min
10		Base teams revisit their own definitions of cooperation and collaboration.	10 - 15 min
11		Invite a business or community leader to speak. **And/Or** interview friends, family, and other community members.	45 min - 1 hr
12	assessment	Create a class puzzle about collaboration.	20 - 30 min
13	assessment	In small groups, read vignettes about people working together and identify if cooperation, collaboration, or both are occurring.	30 - 45 min
14	portfolio	Journal assignment.	10 - 15 min
15	assessment	Create a rubric with some of the collaboration concepts.	15 - 20 min
16		Facilitate the Channels problem-solving activity to practice cooperation and collaboration.	15 - 45 min
17	self assessment	Using the rubric, have each student rate him/herself on collaborative skills.	10 - 15 min

Time Frame: 4.5 - 7 Hours

LESSON 1.C SEQUENCE: DIFFERENTIATE BETWEEN COOPERATION AND COLLABORATION

Create base teams.
Students will spend more time in small groups. Base teams are groups that stay together over time, providing continuity. To create the teams, ask students to secretly write down three to five other people in the class with whom they would like to be on a base team. Collect the cards and create the teams based on who students put on their cards. Tell them that they will be on a base team with **at least one person** they wrote on their cards.

Students will not always work in these teams, but will discuss deeper issues and do larger projects together. It is nice to mix things up a bit so that students are not always with the same people.

Facilitate the Human Treasure Hunt activity using base teams.
Make sure to process how and why they do this activity.

activity

HUMAN TREASURE HUNT

Focus: Getting to know each other, cooperation and competition
Materials: Treasure Hunt sheets, pens/pencils
Time: 15 - 20 minutes
Sequence: Icebreaker
Sources: Laurie Frank learned this from Karl Rohnke. See *Quicksilver* by Karl Rohnke and Steve Butler.

Suggested Procedure

1. Divide the class into small groups of 4 or 5.
2. Pass out copies of the Human Treasure Hunt and explain that this exercise will help them learn more about each other.
3. Then tell them that the object is to get as many points as possible. **It is important how you phrase this because it must leave the door open for them to add up points of other groups, and even the whole class, rather than simply adding the points of their own small group.**
4. Give them about 15 minutes to complete the task. Observe how they behave. Are they outwardly competitive, worrying about others cheating, sharing their answers openly, etc.
5. Accept creative interpretations of the questions.

Sample Processing Questions

- Was this a competitive or cooperative activity? What did you do to compete and/or cooperate?
- Was getting points your only goal in this activity? What were some other goals?
- How did you stretch the rules in order to get more points?
- What were some other ways to get more points?

Facilitation Notes
This icebreaker activity can be used to look at the ideas of competition and cooperation. People generally compete, rather than cooperate or collaborate, in this activity. The rules do not state that they cannot add the numbers of other small groups to their own. Technically the whole class could add their points together so that everyone gets "the most points."

Human Treasure Hunt

Adapted from the activity of the same name by Steve Butler in *Quicksilver*.

The right to pass is in effect. That means you share what you want to with this group of people. It is okay to pass.

Categories: All 1 point + Bonus Points

1. For each different birthday month.
   5 bonus points – born on a holiday
2. For each birth state represented.
   5 bonus – born overseas
3. Add up your shoe sizes.
   5 bonus points – wearing sandals
4. For visiting each of the following: Grand Canyon, Sears Tower, New York City, Niagara Falls, Washington, DC.
   5 bonus points – for three, 7 bonus points – for four, 10 bonus points – all five
5. For riding to school in a car today.
   5 bonus points – if you took the bus, 10 bonus points – if you walked
6. For each parade you have been in.
   7 bonus points – if you played in a marching band
7. For each sibling, living or deceased (includes adopted, step, and half-sibling).
   10 bonus points – for twins
8. Add up the letters in everybody's last name.
   7 bonus points – for each name containing a Z, Q, or X
9. Five points for each language that you speak *fluently*.
   12 bonus points – for three or more
10. For each year you've lived in the place you are living right now.
    5 bonus points – your whole life, 5 bonus points – less than a year
11. For each pet you have right now.
    5 bonus points – for more than 6 pets, 3 bonus points – no pets
12. For each person NOT wearing a watch.
    3 bonus points – no jewelry
13. For each person who can roll their tongue.
    7 bonus points – if you can turn your tongue upside down (in your mouth!)
14. For all those with colored socks (pattern and floral count).
    10 bonus points – no socks

Facilitate the Puzzles problem solving activity that requires cooperation.
Analyze how they did with the initiative. When did they work very well together? What were they doing when they were working well? When were there times they were not working well together? What was going on then?

activity

Puzzles

Focus: Taking turns, group goals, decision making, leadership
Materials: Legos™, eye coverings (optional)
Time: 15 - 20 minutes
Sequence: Problem Solving
Sources: See "Le Cav" in *Funn 'N Games* by Rohnke, and *Journey Toward the Caring Classroom* by Frank.

Suggested Procedure

1. Divide the class into groups of four or five and have each group sit around a table.
2. Give each group a handful of Legos™. Ask them to create a Lego sculpture using no less than 15 and no more than 20 pieces. Allow 10 to 15 minutes for them to create their sculpture.
3. When everyone is done, explain that they have 5 to 10 minutes to create a plan to put the sculpture together again with their eyes closed (or wearing eye coverings). They may not take the sculpture apart during the planning session.
4. When time is up, have them take the sculpture apart, mix up the pieces, and close their eyes.
5. Give them 10 to 15 minutes to put the sculpture back together.
6. When done, give each group an opportunity to share their sculpture with the class and talk about some of the successes and challenges they had with the task.

Sample Processing Questions

- How did you plan to recreate your sculpture? Did it work for you? Why or why not?
- What roles did you take on? Were you the one in there putting things together, or did you wait until your pieces were needed?
- How did you communicate when your eyes were closed? What strategies did you use?
- Which was easier for you—creating the sculpture or re-creating it? What made it easier or harder for you?
- How did you have to cooperate to make this happen?
- When you weren't physically working on the sculpture, were you helping out? (Sometimes not doing something is more helpful than trying to do something.)

Facilitation Notes

Sometimes participants open their eyes before the group is finished. If this should happen, tell that participant that he/she can no longer help the group, but can observe what happens for later discussion.

Complete a large group "Pi Chart" for Cooperation.

A Pi Chart looks like this:

COOPERATION

Looks Like	Sounds Like	Feels Like
• Paying attention • Focusing on the same thing	• Listening to each other • Encouraging words	• Togetherness • Helping

Ask students to come up with descriptors for each area to help identify the essence of cooperation.

Facilitate the ABC Pyramid activity to explore the notion of collaboration.

activity

ABC Pyramid Activity

Focus: Being a team member, cooperation, collaboration
Materials: Alphabet Sheets, pen/paper for each person, flip chart and markers
Time: 20 - 30 minutes
Sequence: Problem Solving
Sources: Laurie Frank learned this activity at a conference sponsored by SolutionTree in 2003.

Suggested Procedure

1. Break the larger group into smaller groups of 5 or 6 people.
2. Make sure each person has a piece of paper and a pen.
3. Tell them that you are going to show them a triangle that contains all the letters of the alphabet, **but for only 10 seconds**.
4. Their task is to memorize as many as possible in those 10 seconds.
5. **They may not write or draw anything while they are being shown the letters, and they may not work with anyone else. This is to be done individually.**
6. Show them the first triangle of letters for 10 seconds.
7. Close it up and indicate that they can start writing.
8. After a moment, show them the letters again and ask them to circle the ones they got correct—to count it as correct, they must have it in the exact place and order as it is shown.
9. On the flip chart, create a chart like this:

Individual	Group	Team

10. Now have each group add up the number of correct letters for each person and divide by the number of people in order to get an average number of correct letters.
11. Ask each group to tell you their average and write it on the flip chart. Make sure you do this in order, so that you can make comparisons later. Generally you will see numbers between 6 and 10. **This is based on individual scores** (Each small group will have an average. This could be a decimal, e.g., 7.2).
12. Next, have them count how many they got correct **as a group**. If anyone in their group got a letter correct, they may count it, but each letter can only be counted once. Numbers usually range between 12 and 16.*
13. Write this number next to their corresponding individual scores. **This is their group score**.
14. Now tell the group that you are going to show them another triangle of letters that looks like the first, but the letters are in a different order.
15. This time they are allowed to plan together how they want to memorize the letters as a group. Show them the **first triangle**, and give them a minute or so to plan.
16. Remind them not to write or draw while you are showing the letters.
17. When everyone is ready, **show the second triangle** of letters for 10 seconds.
18. After 10 seconds close it and indicate that they may now write.
19. After a moment, show them the letters again and ask them to circle the ones they got correct.
20. As before, have them count the number they got correct as a group, remember to only count each letter once.
21. Ask each group to tell you the number they got correct and write it on the flip chart next to their corresponding individual and group numbers. You will generally see numbers between 22 and 26.
22. Look at the numbers and ask the group why they think there was such an improvement. Write their answers on the flip chart. Talk about the difference between working as individuals, as a group, and being a team. What do they think makes a good team? Then connect the idea of teamwork with collaboration.

Sample Processing Questions

- Why do you think working as a group and working as a team produced different results?
- What is the difference between group and team?
- How can we work well as a team?

Facilitation Notes

This activity helps people focus on what really makes for teamwork and collaboration.

* When giving directions, the one that seems most confusing is when counting their group score (#12 above). It seems that the easiest way to count this is to have someone in each group ask who got the first letter. If at least one person raises her/his hand, then it is counted (but only once). Go on to the next letter. If at least one person raises their hand, then it is counted one time. This continues until every letter in the pyramid is addressed. The benefit of doing the group score is that different members in the group can start in a different place (e.g., someone memorizes the letters at the bottom of the pyramid, while someone else focuses on the top).

Alphabet Sheet #1

Y

O N M

U X D A

C R L Z H

K J E G I P

S B Q T F V W

Alphabet Sheet #2

B

R Q P

X A G D

K U O J C

N M H F L S

V T E W I Y Z

Using base teams, answer the processing questions for the ABC Team activity.

1.C sequence 7

Share information about cooperation and collaboration.

Explain that although they are similar, there is a difference between cooperation and collaboration. One way to look at it is:

1+1 = 2 This is Cooperation
1+1 = 3 This is Collaboration

The idea here is that cooperation is simply working together. Following directions is a form of cooperation. Sitting quietly at your desks during this discussion is a form of cooperation. Helping someone with a task is a form of cooperation. Working together (1+1) gets the job done (=2).

Collaboration, on the other hand, allows a group of people to accomplish more than anyone could do on their own or when simply cooperating. Collaboration needs cooperation, but cooperation does not mean collaboration exists. In this case, working together by including everyone and using everyone's strengths (1+1), things happen that cannot necessarily be predicted or expected (=3). Another word for this is "synergy."

Another way to contrast cooperation and collaboration is through the reading of synonyms:

Cooperation: 1 + 1 = 2
concert, joint action, teamwork, co-acting, commonality, concurrence, joining of hands, common effort, joint effort, common enterprise or endeavor

Collaboration: 1 +1 = 3
concur, harmonize, go into partnership with, get together and team up and buddy up, pull together, hold together, hang together, keep together, stand shoulder to shoulder

The differences between the two groups of words is subtle, but powerful. Cooperation is about doing things together, while collaboration implies a relationship between the people while doing things together.

Base teams create definitions for "cooperation" and "collaboration."

Using the information they have gleaned from the activities and discussions, ask them to come up with their own definitions. Each small group should have one definition for each term. This is a time for students to explore their own thinking on these concepts, so there are no wrong answers! Read the definitions out loud and compare similarities and differences.

1.C sequence 9

Read some established definitions of cooperation and collaboration.

Cooperation

"To act or work together for a particular purpose, or to help someone willingly when help is requested."[19]

To cooperate is "To work together toward a common goal or purpose" (Olsen & Pearson, 2000, p. 12.1).[20]

Cooperativeness. "The ability to balance one's own needs with those of others in group activity" (Goleman, 1997, p. 194).[21]

"Cooperation is working together to complete a specific task or achieve a specific goal" (Christ, 2007).

Collaboration

"Relationships that provide opportunities for mutual benefit and results beyond what any single organization or sector could realize alone" (Leader to Leader Institute, 2006).[22]

".... the process of shared creation: two or more individuals with complementary skills interacting to create a shared understanding that none had previously possessed or could have come to on their own" (Schrage, 1990, p. 140).[23]

"...collaborative learning is based upon consensus building through cooperation by group members" (Panitz, 1996).[24]

Collaboration is developing and encouraging relationships in which people can complete a variety of tasks and achieve a variety of goals over an extended period of time (Christ, 2007).

1.C sequence 10

Base teams revisit their own definitions of cooperation and collaboration.
Ask them to modify definitions to reflect new information from the discussion and readings.

Invite a business or community leader in as a guest speaker.
Find someone in the community who is knowledgeable about, and uses, collaboration in her/his workplace or organization. **And/or** interview friends, family, and other community members about their experiences of true collaboration and record notes from the interviews or the situations.

Create a class puzzle about collaboration.
Either have students cut up a large piece of poster board into puzzle pieces, or have it ready for them. Ask students to come up with ways to describe "collaboration." For example, they might say "needs cooperation," or "everyone is an equal." Each concept is written on a puzzle piece. The puzzle is put together one piece at a time and posted on the wall as a reminder of the shared understanding about the nature of collaboration.

In small groups, read the following vignettes about people working together and identify if cooperation, collaboration, or both are occurring.
Using the pi chart on cooperation, the puzzle on collaboration, and their small group definitions, have students decide if there is cooperation, collaboration, or both occurring in each vignette. Ask them to work in small groups to make their decisions. Share thoughts. There are no wrong answers, but it gives you an opportunity to assess if students are thinking about the issues and are headed in the right direction when thinking about these concepts.

Cooperation/Collaboration Vignettes

Vignette #1

Terrence loved to read and had a load of books. He was running out of places to put them, so he decided to build a bookcase. Trouble was, he had never done anything like that before. He decided to asked his friends Jennifer and Craig to help him because they had once built a table. They were happy to help out, even though they had no other experience besides building that one table.

They sat down and came up with a plan. Terrance would do some research about bookcase building, while Jennifer and Craig went to pick up supplies. By the time they returned, Terrance had a drawing and step-by-step directions. Jennifer and Craig had picked up the wood, nails, putty, stain, and varnish. After reading the directions, they gathered the tools they would need: hammer, saw, level, tape measure, pencil, and paintbrush.

Craig was good at figuring out directions, so he read them and helped Terrance and Jennifer figure out what to do. Terrance did most of the building with Jennifer's help. Even though there was a small mistake or two, Terrance ended up with an ideal bookcase to hold his books.

Vignette #2

Sonia and Michele decided to do something they had never done before—go fishing on Lake Superior. This was special because they had never seen such a big lake; it almost looked like the ocean. Both of them had fished but had never been out on open water like this. They traveled almost as far north as possible in Wisconsin to the little town of Bayfield and met Skip, who owned a 30-foot fishing boat. They were excited to learn that they could sleep on board, which meant they could spend a couple of days on the water. It would be like camping without the land around them!

Skip knew the waters in the Chequamegon Bay like the back of his hand. He had grown up there and inherited the fishing business from his father. He checked the weather, and although it looked like some rain might come through, it didn't seem like anything to worry about.

During the day Sonia and Michele learned how to reel in some big fish. They had a couple of good bites, and Sonia caught one very large fish. It was all so exciting. Skip moored the boat near one of the many Apostle Islands around Bayfield. They had a good dinner and were so tired they almost fell into their bunks. The gentle rocking of the boat lulled them to sleep.

A couple of hours later, the gentle rocking turned into swinging. Skip woke the women, saying that the weather had gotten very bad and they had to leave. He would need their help, and they had better put on their raingear. They quickly jumped out of bed and put on their raincoats. When they arrived up top, it was a scary sight. In the dark they couldn't see a lot, but they saw some huge waves. Skip asked them to put on life jackets. He said he couldn't leave the steering wheel once the anchor was pulled up and he needed them to pull up the anchor! He gave them very specific instructions: "When you go out on the deck, don't stand up! You could get pitched overboard." He showed them where the anchor was and how to secure the chain as they pulled.

Sonia and Michele were terrified. They had never done anything like this, but they had to do it in order to get to safe waters. They inched their way out on the wet and slippery deck. The water was frigid as the waves washed over the boat and them. They knew that if they fell into the water, the life jackets would not do much good. Hypothermia would get them in minutes, way before they could be rescued.

They got to the anchor and started pulling. It was easier than they had expected, especially because things are lighter in the water. The boat pitched and rolled, they stayed on their knees, remembering Skip's caution about not standing up. Finally they got the anchor out of the water and near the lip of the boat. It, of course, became instantly heavy, and got stuck under a ridge just below the deck of the boat. They didn't have the strength to haul it up while they were on their knees, and they were in danger of dropping it back into the water. Impulsively, Michele stood up so she could pull the anchor out from under the ridge. Just then a wave hit the boat and she felt herself falling over the side. Skip was yelling, "Get down, get down," but it was already too late. Just as she saw the water coming up at her, she felt herself being pulled back in. Sonia had the back of her life jacket with two hands and somehow found the strength to pull both her and the anchor onto the deck.

They both lay on the deck gasping for air. When they caught their breath, they secured the anchor and made their way back to the cabin. Skip didn't say a word because he was already concentrating on moving the boat out of harm's way. All Sonia and Michele could do was wedge themselves into a corner so they wouldn't tumble around too much and try to forget how sick they were beginning to feel.

Three hours later they were in a safe port, sharing coffee, and recounting their harrowing experience.

Vignette #3

Greg had lived most of his life in a small town where everyone knew everyone else. Now it was time to move on, and he had found a job in Chicago. Going from a town of a couple thousand to a city of millions would be quite an adjustment.

At first the excitement of the move stayed with him. He had so much to do to fix up his new apartment. After awhile, though, the newness started to wear off and he felt very isolated and alone. He was used to people stopping to talk. In Chicago, most people avoided eye contact. It was weird, but even though he was surrounded by people, he had never felt more alone. Even people at work just did their job and went home, leaving him to trudge home alone.

Finally, Greg had enough of feeling sorry for himself. He decided to take the initiative to seek out other people, but he just didn't feel good about going out by himself. One night he called home and confessed to his mother that he was feeling lonely and didn't know what to do about it. She said matter-of-factly, "Well dear, why don't you volunteer somewhere? People always need help."

It was almost too simple. Of course he could volunteer, but where to start? He looked up the address for the Chicago Department of Human Services and visited them after work. They asked about his interests and then put him in touch with a Boys and Girls Club near his apartment. The club was happy to see a volunteer show up on their doorstep! They put him to work right away helping with their after-school clubs.

Greg continued to work at the center, helping with meals programs, supervising recreation for kids, and even raising needed funds. At the end of the year they honored him as their volunteer of the year. As he looked around the room when he got the award, he realized how many friends he had made in such a short time. He finally felt at home.

1.C sequence 14

Journal assignment.
Using their base team definitions of collaboration have each student rewrite the definition to create a personal definition of collaboration.

Create a rubric with some of the collaboration concepts.
As a class, choose five of the collaboration concepts from the puzzle and create a rubric to assess collaborative skills. For example:

Criteria	0	1	2	3
Needs cooperation	Only thinks about self	Doesn't do anything to help, but doesn't hurt process either	Only helps when asked	Takes initiative to help
Everyone is an equal	Bosses people around	Will work with some people but not with others	Does most of the talking, but will listen sometimes	Asks people what they think, listens to various points of view, and takes turns

It may be necessary to break into small groups to work on each of the criteria.

Do the Channels problem solving activity to practice cooperation and collaboration.
After completing the activity and subsequent processing, give each student a copy of the rubric the class created, and have each student rate him or herself on collaboration skills.

activity

CHANNELS

Focus: Cooperative problem solving
Materials: Channels made with 1/2-inch PVC pipe, ball bearing or marble to fit in channel, tin can
Time: 15 - 45 minutes
Sequence: Problem Solving
Sources: Laurie Frank learned this activity from Dan Creely. See *Adventure Education for the Classroom Community* by Frank & Panico, "Pipeline" in *Affordable Portables* by Cavert, "Gutterball" in *Games for Group I* by Cavert, *Journey Toward the Caring Classroom* by Frank, and "Marble Tube" in *Teamwork and Teamplay* by Cain and Joliff.

Channels are made with 1/2 inch PVC pipe, cut lengthwise with a band saw, in 12- to 18-inch sections. You can also get corner molding from a hardware store and cut them into 12- to 18-inch sections. You will also need at least one ball bearing or marble that will fit in the channels and a tin can. In a pinch you can give everyone an 8.5 by 11 piece of paper, and have them fold it lengthwise three or four times until a "V" is formed.

Suggested Procedure

1. Every participant is given a channel. The task is to move the ball bearing across a predetermined area and into the can. It is important to make the distance to travel longer than the group can get to by standing next to each other. The rules are:
   - No one may touch the ball bearing with his/her skin or clothing.
   - The ball bearing may not touch the floor.
   - If either of the above happens, the group must start over.
   - When an individual has the ball bearing in his/her channel, he/she may not walk.
   - Channels may not be made into tunnels by putting one channel on top of another
   - Each person must remain in possession of his/her own channel.
2. Observe how the students plan, listen to each other, and work together. Are there one or a few students doing most of the talking? Are some students left out of the planning? Is someone trying to be heard, but ignored? Does someone come up with an idea that is immediately discounted? Do they stop and come up with a plan, or do they immediately dive in and do the trial-and-error approach? Does everyone seem content with the process or are some appearing to be irritated or frustrated?
3. If the task seems easy and they finish quickly, ask them to try it again, but this time the ball bearing or marble may not move backwards at any time. If it rolls backwards, they must start over.

Sample Processing Questions

- How did you decide to organize your group in order to accomplish this task? Why not spread out more?
- Given the nature of the task, could this have been accomplished alone?
- How was this task accomplished in a group, when you could not have done this easily alone?
- How was each person a leader and a follower in this activity?
- What strategies worked for you? What did not work?

Facilitation Notes

This task is harder than it appears, and a group must have the capacity for patience. It is common for the ball bearing to fall more than once, with the group having to start over many times; thus, the frustration level can rise. That is usually when communication breaks down and people stop working together. When this happens, it provides good fodder for discussion—focus on how to recognize the need for, and provide structure for, a group when facing a difficult problem.

1.C sequence

Using the rubric, have each student rate him/herself on collaborative skills.

Have each student use the rubric to assess his/her own collaboration during the Channels activity. After sharing it with the teacher, it can be placed in the portfolio.

Lesson 1.D Overview: Use a Variety of Skills and Tools for Collaboration

Focus: Encourage open communication
Encourage participation
Practice reflection

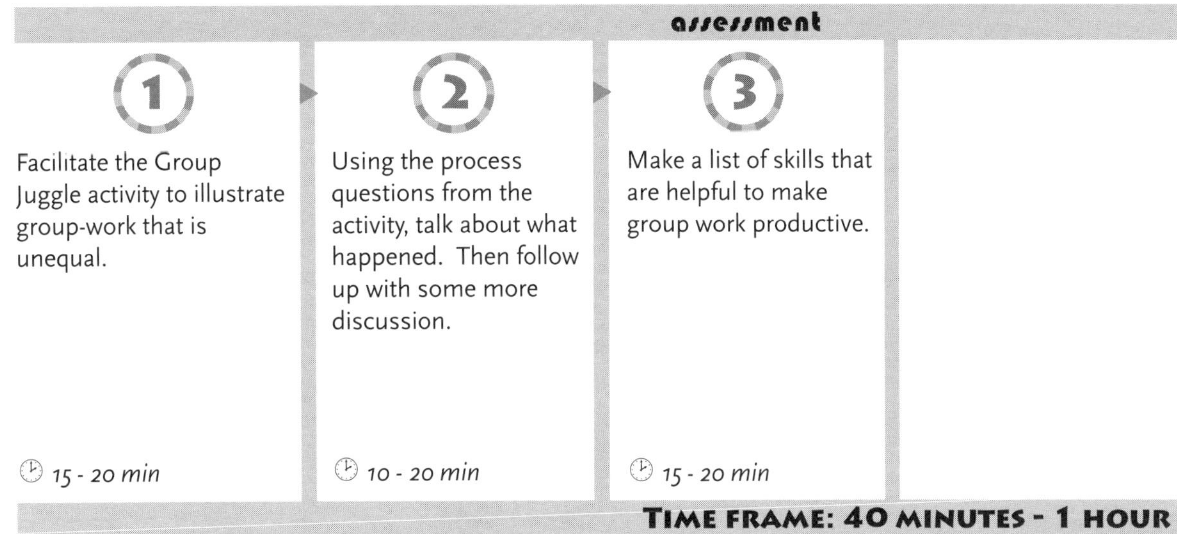

Lesson 1.D Sequence: Use a Variety of Skills and Tools for Collaboration

Facilitate the Group Juggle with a Twist activity to illustrate group-work that is unequal.

activity

Group Juggle with a Twist

Focus: Decision making, leadership, taking turns, group goals
Materials: Many soft throwable objects (wadded up pieces of paper work in a pinch)
Time: 20 - 40 minutes
Sequence: Problem solving
Sources: See "Group Juggling" in *More New Games* by Fluegelman, "Group Anger Juggle" in *Adventures in Peacemaking* by Kreidler & Furlong, "Juggling for Our Lives" and "Low-Drop Juggle" in *Adventure Education for the Classroom Community* by Frank & Panico, *Changing the Message* by Albin, *Diversity in Action* by Chappelle & Bigman, "Messages" in *Games for Group I* by Cavert, "Basic Group Juggle and Variations" in *Journey Toward the Caring Classroom* by Frank, *Quicksilver* by Rohnke & Butler, and "Community Juggling" in *Teamwork and Teamplay* by Cain & Jolliff.

Suggested Procedure

1. Clear the desks or tables away and stand in a circle.
2. Tell everyone they need to know the name of the person on their right. Give them time to see who that is.
3. Take one item and toss it around the circle to the right. Each person calls the name of the person to his/her right before throwing them the object. Go around a couple of times.
4. Now stop and place the objects on the ground. Say "This is 'A' as in apple" and have everyone line up in alphabetical order, re-forming the circle.
5. Make sure everyone is in order by having each person call out his/her name.
6. Ask each person to identify where the person who used to be on their right is now. They will continue to throw to that person.
7. Pick up an object and begin the pattern by calling the name of and throwing to the person who used to be on your right.
8. Now tell them that, as a group, they will juggle all of the objects. Everyone always throws to the same person.
9. Then throw the objects, one at a time. When an object returns to you set it down, until all objects make it back to you.
10. Try the juggle one more time.
11. This time, though, there is a twist. After throwing the first two items, give the items to the person next to you to throw to their same person. Choose the person on your right or left who gets the object later in the process (i.e., if the person on your left is the third person to throw, and the person on your right is the eighth person to throw, then give it to the person on your right). Continue putting the objects down when they get to you.
12. Some of the people in your group will be surprised that items stop coming to them. Others will not even notice, especially if they continue to be in the group that throws.

Sample Processing Questions

- What skills/qualities did you need in order to successfully juggle all of the objects?
- What made the difference between the first, second, and third attempts?
- How did it feel for those of you who were not included in the final attempt?
- How was it for those who were included?

Facilitation Notes

By having the person next to you start the process, it automatically excludes those who got the object before that person. In other words—some of the people are doing all of the work, while others are not involved. This situation opens the door for some processing around group work and an age-old problems—who does the work? Ideally everyone shares the load. Sometimes, however, it seems that one or a couple of people are doing most of the work. When this happens, it is important to ascertain whether it is because people are choosing not to work or whether the few people working are excluding the others. Many times, of course, it is a combination rather than an either/or situation.

Using the processing questions from the activity, talk about what happened. Then follow up with some more discussion.

How did it feel to those who were doing most of the work while others were doing nothing? Was it of their choosing? How did it feel to those who had nothing to do? Was it of their own choosing?

Address the age-old dilemma of group work—sharing of responsibility and workload. Talk about the need to discover a balance.

Make a list of skills that are helpful to make group work productive.

Solicit ideas from the students about what skills are needed to make group work more productive. Include skills that can help deal with conflicts when they arise, such as when someone feels that he/she is doing most of the work, or when someone is feeling excluded. Tell the students that the rest of this lesson is going to be about exploring communication, decision-making, and conflict-resolution skills, so that they can develop them for subsequent group work.

Lesson 1.D[1] Overview: Communication Strategies

Focus: Encourage open communication
Listen actively
Check for understanding
Promote active listening
Attend to non-verbal cues

Active listening: Facilitate the A What? activity.

10 - 15 *min*

2

Using the processing questions from the activity, discuss what strategies people used to pay attention, even when it was chaotic.

10 - 15 *min*

3

Present guidelines for active listening.

5 - 10 *min*

4

Facilitate the 1-1 Interview activity to practice active listening.

15 - 20 *min*

self assessment

Using the supplied rubric, students rate their active listening skills from the 1-1 Interview activity and their general active listening skills.

10 - 15 *min*

6

Non-verbal cues: Facilitate the Interactive Video activity.

10 - 15 *min*

7

Using the processing questions from the Interactive Video activity, discuss what people keyed into during the activity.

5 - 10 *min*

8

Present a list of body-language cues.

10 - 15 *min*

assessment

Participate in feelings role plays to identify unspoken messages.

15 - 20 *min*

Time frame: 1.5 - 2.5 hours

LESSON 1.D¹ SEQUENCE: USE A VARIETY OF SKILLS AND TOOLS FOR COLLABORATION: COMMUNICATION STRATEGIES

Active Listening: Facilitate the A What? activity.

1.D¹ sequence

activity

A WHAT?

Focus: Active listening, taking turns in conversation
Materials: None
Time: 10 - 15 minutes
Sequence: Deinhibitizer
Sources: *More New Games!* by Fluegelman, *The Incredible Indoor Games Book* by Gregson.

Suggested Procedure

1. Sit in a circle.
2. Start passing an object around by turning to the person on your right, handing them the object, and saying, "This is an aardvark."
3. They respond by saying, "A What?" Then you say, "An Aardvark."
4. They turn to the third person and say, "This is an Aardvark."
5. When the third person says, "A What?" the second person turns back to you and says "A What?" and you say, "An Aardvark!"
6. They then pass "An Aardvark" to the next person.
7. This pattern continues.
8. Next start a different object going to the left: "This is a gorilla." "A What?" "A Gorilla." And so on....

Sample Processing Questions

- How did it feel to get caught between the aardvark and the gorilla? Did it help to have people make suggestions to you? Why or why not?
- Everything was going so smoothly, what caused the problem?
- Was it possible to hear what was going on all the time during this activity?
- Why is it important to take turns when talking in a conversation or group discussion?
- How can we make sure people have a chance to be heard during discussions in this class?

Facilitation Notes

This activity causes confusion. Everything works out pretty well until the aardvark and gorilla meet. Suddenly people don't know who to listen to or who to tell what to. Some groups figure it out, but that is not the norm. Communication breaks down very rapidly, and the aardvark and gorilla rarely make it all the way around the circle. This provides a very good opportunity to process about communication and taking turns in conversation.

Make sure that students are ready to be able to make mistakes before attempting this activity. Sometimes the person (or people) who get caught between the aardvark and gorilla get very confused and others start yelling at them (it is always much easier to see the problem/solution from the outside).

1.D[1] sequence 2

Using the processing questions from the activity, discuss what strategies people used to pay attention, even when it was chaotic.

Ask how they were able to make sense of what was going on. When things were crazy, what did they try to do to maintain focus? These ideas set the stage for teaching active-listening strategies.

1.D[1] sequence 3

Present guidelines for active listening.

- Look at the person who is talking.
- Focus on what the person is saying.
- Show that you are listening by periodically nodding, or saying something like "uh, huh."
- Ask clarifying questions, or repeat back what you heard, if necessary.

1.D[1] sequence 4

Facilitate the 1-1 Interview activity.

This activity lets people consciously practice their active listening skills.

activity

1-1 Interview

Focus: Active listening, taking turns in conversation
Materials: None
Time: 15 - 20 minutes
Sequence: Ice Breaker
Sources: Unknown

Suggested Procedure

1. Everyone gets a partner and is given a set period of time to learn everything they can about their partner.
2. This can range from one minute and up, depending upon how much time you have or how much pressure you wish to add to the situation.
3. After the allotted period of time, each person introduces his/her partner to the group.

Sample Processing Questions

- What did you learn about your partner?
- How did you use your active listening skills? Did it feel natural?
- What strategies did you use to remember the information?

Facilitation Notes

Many people feel it is easier to introduce someone else other than him/herself, since it takes the focus off of them. This activity offers that opportunity. It may also be necessary with some groups to establish some ground rules while people are speaking, since some people have a difficult time speaking in front of a large group. If someone decides not to introduce their partner due to nervousness, try asking that person questions to help them focus. Their partner can also introduce him/herself.

Using the supplied rubric, have students rate their active listening skills from the 1-Interview activity and their general active listening skills.

After the 1-1 Interview, have students fill out the following rubric to indicate how they did with active listening skills, then share it with their interview partner. Have students reflect on their active listening skills in general.

Criteria	1	2	3
Eye contact	Eyes wandering. Rarely looks at the person.	Look at the person who is talking, but looks away when he/she is looking at you.	Make eye contact with the person who is talking, if culturally appropriate, and without staring.
Focusing on conversation	Thinking of other things or listening to someone else's conversation.	Fading in and out. Not hearing all the conversation.	Totally focused and engaged.
Showing that you are listening	Staring blankly.	Looking at person and maybe nodding every once in awhile.	Nodding head or making sounds that indicate you are hearing what the other person is saying.
Indicating interest and understanding	Is disinterested and does nothing.	Asks a question or makes a comment.	Shows curiosity by asking questions about what the person said for more information, or repeats something back that was interesting.

.D1 sequence 6

Non-verbal cues: Facilitate the Interactive Video activity.

activity

Interactive Video

Focus: Active listening, hidden agendas, perspective taking, non-verbal cues
Materials: None
Time: 10 - 15 minutes
Sequence: Ice breaker
Sources: See *Journey Toward the Caring Classroom* by Frank, and "Captain Video" in *New Games for the Whole Family* by LeFevre.

Suggested Procedure

1. Ask students to sit in a circle with everyone facing away from the middle.
2. One person stands in the middle, taps another on the shoulder so that he or she can watch, and does a simple visual routine (e.g., hands on hips, tapping toe, with a cheerleading jump at the end).
3. Person #2 taps the next person and repeats the routine
4. Person #3 continues by tapping #4, etc.
5. Once someone has done the routine, s/he may watch the progression, but may not comment on it (other than to laugh when appropriate).
6. When the routine goes all the way around the circle, then Person #1 and the last person face each other. They count to three and do their respective routines (#1 does what he did. The last person does what she saw).

Sample Processing Questions

- How did the message change over time?
- What do you think was the cause of the changes?
- Was it possible to take in the whole message? How?
- What did you key into when trying to get the message that was being communicated?
- How can our own perspectives or abilities cause us to change a message?

Facilitation Notes

As you can see, this is a visual form of the game Operator. Generally, the routines change by becoming simpler, as it is impossible to keep track of all the details that one is witnessing. There is usually much laughing as the audience sees the changes taking place.

With some students, especially in the elementary grades, they sometimes do not wait for the person to turn around before beginning the routine. Obviously, that person will not have the benefit of seeing the full routine and will be at a severe disadvantage. This is a good teachable moment to focus on how one must be tuned in to the person with whom they are trying to communicate.

Using the processing questions from the Interactive Video activity, discuss what people keyed into during the activity.
Note the idea that there is much more to a communication than what is being said. Body language is a powerful communication tool.

Present a list of body-language cues.

It has been said that 93% of communication is nonverbal. This means that body language and tone of voice are vital when trying to communicate. What does this say about e-mail?

David Givens from the Center for Nonverbal Studies (CNS) has compiled a list of nonverbal communication cues. Please note that none of these are universal. There are always exceptions, and interpreting body language has to happen in context. For example, crossed arms can mean significantly different things depending on the mood of the person and the environment.[25]

Some Body-Language Cues

Arm-Cross: Arm-cross can simply be a comfortable, relaxed position. When pulled in tight, it can signify nervousness. It can also be a way to protect oneself, say in a crowded room.

Blank Face: Blank face can mean "leave me alone." A blank face is a way to keep people away. If someone is in an airport, for example, and doesn't want to talk to the people around them, she/he may have a blank expression so as not to invite conversation.

Body Alignment: Body alignment is leaning toward someone, which can show loyalty or respect. Sometimes, in a meeting, it is possible to figure out who is the boss by the number of people leaning in his or her direction.

Eye Contact: Eye contact can be very emotional. Sustained eye contact causes people to become uncomfortable. It is also cultural. In the United States, people are taught to make eye contact when someone is speaking. In Japan, listeners are taught to avoid making eye contact when someone is speaking.

Hands Behind Head: Hands behind head can mean that a person disagrees, is uncertain, frustrated, or angry. It is generally a sign of negative feelings.

Hands on Hips: Hands on hips can indicate that a person is ready to do something—ready to act, perform, get started.

Lip-Purse: Lip-purse can denote disagreement.

Palm Up: Palm up suggests that the speaker is nonaggressive, as in offering a peace sign. It enlists the listener as an ally.

Palm Down: Palm down suggests dominance, assertiveness, confidence. It emphasizes speaking points, and can make them seem more convincing.

Shoulder Shrug: Shoulder shrug generally means the person is uncertain. Even if someone says, "I'm sure," saying it with a shoulder shrug can mean that they really aren't sure. It can mean "I don't know."

Tone of Voice: Tone of voice can change the meaning of what is said, and reflects the mood of the speaker. For example, consider the meaning of "that's a nice shirt you have on" when you really mean it and "that's a nice shirt you have on" when you are being sarcastic and really mean the opposite of what your words say.

Participate in feelings role-plays to identify unspoken messages.

Break into small groups. Write each feeling word from the following list on an index card. Give each small group three of the cards. Their task is to create a role-play using the feelings involved and some of the body-language cues that have been shared, as well as other types of body language that is not on the list that was presented. After each role-play is presented to the class, discuss the body language that people saw, and see if they can guess what three feelings they were attempting to portray.

Encourage students to show dissonance at some point in the role-play by saying one thing and showing body language that conflicts with what is being said.

Feelings Words

Apathetic
Bold
Bored
Brave
Calm
Determined
Disappointed
Embarrassed
Excited
Friendly
Frustrated
Guilty
Helpful
Helpless
Hopeful
Lazy
Nervous
Optimistic
Powerful
Proud
Quarrelsome
Relaxed
Restless
Sad
Shy
Surprised
Tough
Upset
Vulnerable
Worried

LESSON 1.D² OVERVIEW: DECISION-MAKING STRATEGIES

FOCUS:
- Encourage diversity of opinion
- Discourage groupthink
- Facilitate brainstorming
- Facilitate collaborative decision-making
- Foster creativity
- Seek consensus about group decisions
- Promote compromise

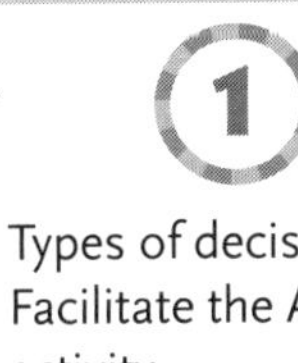 1. Types of decisions: Facilitate the All Toss activity. 15 - 30 min	2. Use the processing questions included with All Toss to explore how decisions were made. 5 - 15 min	 3. Introduce the concept that group decisions are not all alike, and present a decision-making strategy list. 15 - 20 min	 4. Using the strategy list, facilitate the Don't Spill the Beans activity. 15 - 20 min	 5. Using the processing questions from the Don't Spill the Beans activity, discuss their decision-making strategies. 5 - 15 min
	portfolio		portfolio	
 6. **Brainstorming:** Facilitate the SCAMPER activity. 10 - 15 min	 7. Facilitate the Creativity: Four Measures of Style activity. 30 - 60 min	 8. Hold a small or large group discussion about brainstorming. 10 - 15 min	 9. Facilitate Your Personal Network brainstorming activity. 20 - 40 min	 10. **Consensus:** Facilitate the Pathfinder activity. 15 - 30 min
assessment				assessment
 11. Journal assignment. 10 - 15 min	12. Teach consensus guidelines and Fist to Five strategies. 10 - 20 min	 13. Facilitate the Batten Down the Hatches Activity. 20 - 45 min	14. **Decision-Making Process:** Introduce the STEPS decision-making process. 10 - 15 min	 15. In small groups use the STEPS process to arrive at solutions to a written problem. 10 - 15 min
	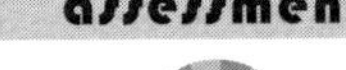 assessment	self assessment		
16. Read about Decision-Making Pitfalls. 20 - 30 min	17. In small groups, return to the Shackleton and *Into Thin Air* articles to find examples of decision making. 30 - 40 min	 18. Journal assignment. 5 - 10 min		

TIME FRAME: 4 - 8 HOURS

LESSON 1.D² SEQUENCE: USE A VARIETY OF SKILLS AND TOOLS FOR COLLABORATION: DECISION-MAKING STRATEGIES

Types of decisions: Facilitate the All Toss activity.

activity

ALL TOSS

Focus: Taking turns, group goals, decision making
Materials: A soft throwable object for each person
Time: 15 - 30 minutes
Sequence: Problem Solving
Sources: See "All Catch" in *Adventure Education for the Classroom Community* by Frank & Panico, *Journey Toward the Caring Classroom* by Frank, and "Upchuck or Barf Ball" in *Quicksilver* by Rohnke & Butler.

Suggested Procedure

1. Clear the desks or tables away and stand in a circle.
2. Give each person an object and ask him to put it at his feet.
3. Tell the group that the object of this activity is to see how many items can be thrown and caught all at the same time. All objects must be thrown at the same time. They cannot be thrown to oneself or to the person on either side of the thrower.
4. Start with one object. Count to three, and throw your object in the air. If it is not caught, then try it again until it is.
5. When the object is caught, ask someone else to pick up his or her object, also. They count to three and both throw their objects. If both are caught, pick up a third object.
6. Anytime an object is dropped, it is taken out for that round and whatever was caught is thrown again. For example, if five objects are thrown and two dropped, the next round involves the three that remained in play. If those are caught, then a fourth one is picked up for the round after that.
7. As the task becomes more difficult, allow time for the students to create strategies.
8. If time is running low, or if frustration or boredom begins to set in, then ask the class to set a goal for how many attempts they wish to try to throw all of the items at the same time.

Sample Processing Questions

- When did you decide to really begin to communicate? What made you decide to do that?
- Why did your strategies change?
- Who made the decisions about what to try and how to do it? Did someone give an idea and everyone agreed, or did someone give an idea and no one disagreed? (These are two different things.)
- How were decisions made? What was your process for decisions making?
- What if the stakes were higher (e.g., dropping them would cause you to fail the class, or catching them all would cause you to get a job or earn money)? How might your decision-making process change?

Facilitation Notes

Because this is a simulation activity the stakes are low, which has an effect on how people make decisions. One way to illustrate a high-stakes situation is to head outside and add in one or two raw eggs as throwables. Note how decision-making changes. Many times things slow down and people create a conscious plan to take care of the high-stakes situation (the egg).

At first, most groups do little planning when attempting this task. As more items are added, however, and items begin to hit each other, then the need arises to communicate and collaborate more closely. Although it is unstated, many groups decide to stay in a circle to complete this task. Some groups, though, decide to get into two lines and throw across from each other, especially when items begin to hit each other in the middle of the circle.

Some of the more interesting conversations after this activity surround the issues of goals. Is it more important to be organized, or more important to have fun? Can one be organized and still have fun?

With a large class, try dividing into two groups for this activity. Once both groups have had a chance to try this, then combine the two into one large group for a bigger challenge. For an even bigger challenge, add the rule that every time an item hits the floor, they must totally start over.

Use the processing questions included with All Toss to explore how decisions were made.

1.D^2 sequence 3

Introduce the concept that group decisions are not all alike, and that there are different ways to make decisions based on the situation that is presenting itself.

Different Types of Decision-Making Strategies

Informal and formal decision-making: Decision-making falls into two general categories: informal and formal. It is important for group members to understand the types of decisions they are making, and choose decision-making strategies depending on their goals. Informal decision-making usually involves daily decisions and items that are not related to the structure of an organization or to the process by which organizations establish their governing rules.

Autocratic decision-making: Autocratic decisions are made by one person or a small group of people who are not necessarily affected by the decision. This method is used in emergency situations, as when a building needs to be evacuated. It is also used when decisions don't affect the governing structure of a group or organization, such as setting the schedule for an event.

Participative: Participative decisions are made by the group or team affected by the decision, and may be made by a majority vote or consensus of the group. Participation is encouraged when high commitment is needed and when the decision calls for a variety of judgments. Effectively designing a community-service project involves participative decision-making.

Avoidance: Avoidance in decision-making, or putting the problem on the back burner, is used when the problem will go away, is determined not worth the effort, or may not be solvable by the assembled group. Avoidance might be used effectively when some members of the student body want to tell the administration when to schedule make-up days.

Compromise: Compromise in decision-making means that all members accept the minimum with which they can live, and it is used when time is short and the consequences of the decision are not severe. Compromise is a participative process. Selection of a theme for Homecoming could be accomplished effectively through compromise.

Majority decision-making: Majorities are determined by voting, with a simple majority being one vote more than half of the votes cast, a 2/3 majority being 2/3 of the votes cast, and a 3/4 majority being 3/4 of the votes cast. It is a participative process. Decisions that alter the structure of the group or the procedures for decision-making usually require a higher percentage of the votes cast, indicating the significance of the issue. Election of officers is usually by simple majority. Changing the constitution, ending the filibuster, limiting debate, and sometimes spending the money requires higher vote margins.

Consensus decision-making: Consensus requires agreement of all the members of the group, and is used when a common goal is established, there is ample time, and there is or needs to be a high commitment to the group's vision or task. Consensus can be viewed as the best decision for the group at this time, having heard and considered all ideas from all involved. Consensus is a participative process best used for major decisions, such as selection of a year-long project or whether or not to move a business to a different location. Consensus does not require absolute agreement among all members, but it does mean that all members agree to live with the decision even if some members would prefer a different decision.

1.D[2] sequence 4

Using the strategy list, facilitate the Don't Spill the Beans activity.
Begin the activity by dividing the class into groups of 3 to 4. Ask each group to discuss the different decision-making strategies and choose one to focus on. Acknowledge that many of the other strategies may come into play, but that they need to use their chosen one at least once during the activity.

activity

Don't Spill the Beans

Focus: Decision making
Materials: Bandana for each group of 3 or 4, a cup for each group, lots of dry beans, broom
Time: 15 - 20 minutes
Sequence: Problem Solving
Sources: See *Games for Teachers* by Cavert & Frank.

Suggested Procedure

1. Give each group a cup and a bandanna. Have the beans in a container in the middle.
2. Tell them that their task is to transport a cup of beans from point A to point B.
3. Ask each group to designate their own point A and point B. It will be different for each group and it doesn't matter how far apart they are. Give them a minute to make their decision.
4. Inform them that they may change their point A and point B later if they find it too challenging or too easy.
5. Tell them again that their task is to move a cup of beans from one place to another using the following rules:
   a. The cup of beans must be placed on or in the bandanna.
   b. Everybody has to be involved in transporting the beans.
   c. If any of the beans spill (even into the bandanna), they must go back to point A.
   d. They can choose any way to carry the beans, as long as they are using the bandanna.
6. Once they reach point B, they are to go back to point A and try it again, BUT they must create a more challenging way to transport the beans than they did the last time.
7. Give them about 10 minutes to try as many ways to transport the beans as possible.
8. Have a broom ready to clean up any beans that may be spilled.
9. When done, ask them to write down all the decision-making strategies used.

Sample Processing Questions

- Who made the decisions about where point A and point B were? Did one person make a suggestion and all agree or did one person make a suggestion and no one disagreed? (These are two different things). Did you talk about different ideas and choose one? What are other ways the decision was made?
- How did you choose which decision-making strategy to use? How did you decide how to decide?
- How did the decision-making strategy you chose play out? Did it work or not work for you? In what ways?
- Were your decisions mainly formal or informal? How could you tell?
- When collaborating, why might it be important to be aware of how decisions are made?

Facilitation Notes

This activity can take time, and different groups become engaged at different levels. Some small groups seem to tire of it quickly, while others continue to be creative and fascinated with it for a while. The level of group involvement will determine how much time you allow. If they are engaged, give them more time to experiment. If interest wanes, call them back together. It may be necessary to float around and encourage some groups to push their challenge level up a bit. Many times the "fear of failure" keeps groups at a level that is easy. This can lead to boredom.

Using the processing questions from the Don't Spill the Beans activity, discuss decision-making strategies.

1.D^2 sequence 5

Brainstorming: Facilitate the Scamper activity.

Brainstorming is a useful tool when making decision, solving problems, or dealing with conflict.

1.D^2 sequence 6

SCAMPER

Focus: Brainstorming, creativity
Materials: Paper and pencil for each group
Time: 10 - 15 minutes
Sequence: Problem Solving
Sources: Laurie Frank learned this activity from Dick Prouty in the early 1980's.

Suggested Procedure

1. Divide the class into groups of 3 to 5 students.
2. Make sure each group has a piece of writing paper and a pencil.
3. Have them choose someone from their small group to be the scribe.
4. Ask everyone to dig into his/her pockets/bags and take out an item of interest. It can be anything (i.e., keys, pictures, books, etc.).
5. Ask each group to look at all the items and choose one that seems the most interesting to them.
6. Tell them that they are going to do a brainstorming activity. They will have three minutes to brainstorm as many uses for their item as possible.
7. Remind them of brainstorming rules:
   - Accept any and all suggestions.
   - No discussion. No judgments. Simply write down the suggestions.

• Quantity is the goal—get as many ideas out there as possible.
• Look for opportunities to "piggyback" (i.e., take someone's idea and expand on it).
8. After the three minutes are up, have each group choose 3 or 4 items to share with the class.

Sample Processing Questions
- How were the brainstorming rules helpful?
- Did you end up with ideas that were surprising to you? How did that happen?
- Look at your lists. Do you see instances of piggybacking? Share some.

Facilitation Notes
Brainstorming seems simple to those who have done it many times; however, it is an important skill for group work, for conflict resolution, and problem solving. It is especially important to emphasize the nonjudgmental nature of brainstorming because it encourages creativity and quantity. The idea, of course, is to get the stuff out there and make judgments about it later. Sometimes the most outlandish idea can be modified and end up being the best option.

1.D[2]
sequence

Facilitate the Creativity: Four Measures of Style activity.
This activity lets students practice brainstorming while assessing their creativity. After doing the activity, have students write insights in their journal. Save this in their portfolios.

activity

Creativity: Four Measures of Style[26]

Focus: Brainstorming, creativity
Materials: Creativity sheet for each person
Time: 30 - 60 minutes
Sequence: Problem Solving
Sources: Badger High School Leadership Manual

Suggested Procedure
1. People do this individually.
2. Read the information on the sheet.
3. Explain that creativity can be measured in many different ways.
4. Have everyone tally their own scores.
5. For each score, have students line up according to their number.
6. Note how people have different strengths, which are useful to know when working together.

Sample Processing Questions
- What is something you learned about your own creativity?
- How did brainstorming these things teach you something about yourself?
- How was brainstorming in a group different than brainstorming individually?

Facilitation Notes
Be alert for students who may show less creativity than others, and may put themselves down because of it. Note that it is important to have a variety of strengths within a group. In this case it's creativity. In another case we may need verbal ability, or flexibility, etc.

CREATIVITY: FOUR MEASURES OF STYLE

Creativity is a mysterious trait. Experts keep coming up with different ways to define and measure it. Dr.·E. Paul Torrence, an expert in creativity, has broken down the creative thinking process into these four characteristics:

1. Fluency: number of ideas you can generate
2. Flexibility: different kinds of ideas you can generate
3. Originality: do you own or borrow your ideas
4. Elaboration: amount of detail in your ideas

Use these four characteristics to find out how creative you are. You may surprise yourself.

Directions: This is a simple pattern found over and over again in nature. Take the design and embellish it so that it functions in a different way. What do you think it is? What do you think it might be? Come up with as many ideas as you can. Be brave—be crazy! There are no right or wrong answers. You have five minutes.

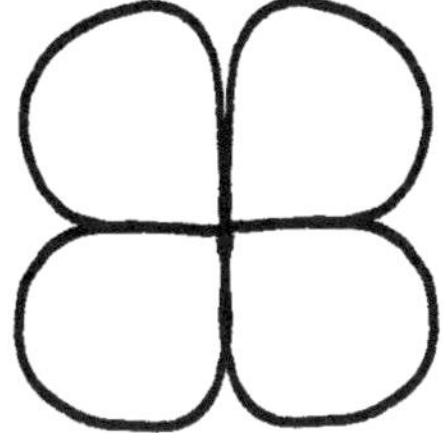

How might this function....

A. In a Kitchen

____

____

____

____

____

____

B. In a Backyard

____

____

____

____

____

____

C. On a Beach

____

____

____

____

____

____

D. On an Automobile

____

____

____

____

____

____

E. In a Library

____

____

____

____

____

____

F. On a Mountain Top

____

____

____

____

____

____

Style Analysis... Creativity

Fluency: How many ideas did you come up with for each drawing?

A: B: C: D: E: F:

Total your points to come up with your **fluency score:** ____________

Flexibility: How many different kinds of ideas did you come up with? (Think of each kind as a category: clothes or furniture or animals. If you identified two animals, give yourself one point because they are in the same category).

A: B: C: D: E: F:

Total your points to come up with your **flexibility score:** ____________

Originality: Are your ideas borrowed? (Compare your ideas with the ones listed below, and cross out anything you wrote that is the same or very similar with these).

A: potato masher, sponge, heating pad
B: garden tool, flower, leaf
C: umbrella, raft, flower
D: air freshener, fan, dashboard, air bag, cup holder
E: fan, study seat, bookmark
F: flower, mountain top, four leaf clover, rock

Total your points to come up with your **originality score:** ____________

Elaboration: How detailed are your ideas? (Give yourself one point for each adjective you used; one point for each adverb you used; one point for each verb. For example: sombrero on table gets no points; sombrero sitting on table gets one point (for using the verb "sitting").

A: B: C: D: E: F:

Total your points to come up with your **elaboration score:** ____________

Now add up your scores to find your Grand Total:

Fluency score:

Flexibility score:

Originality score:

Elaboration score:

Grand total Creativity Score: ________________

Hold a small or large group discussion about brainstorming.

When is brainstorming useful or not useful? How can it be used in a group to help the process along? When are there times when brainstorming can actually get in the way of getting things done?

Facilitate Your Personal Network brainstorm activity.

This is one more chance to practice brainstorming, while combining it with another purpose that will be useful to the student. When done, put these in their portfolios for later use.

activity

Your Personal Network[27]

Focus: Decision making
Materials: Flip chart paper and markers, stopwatch, personal inventory sheet
Time: 20 - 40 minutes
Sequence: Problem Solving
Sources: Badger High School Leadership Manual

Suggested Procedure

1. Prepare by setting up four flip chart papers and markers around the room. Each one with a different label: Family, Community, School, Other.
2. Divide the class into four groups, and have each group start at a flip chart paper.
3. Their task is to brainstorm for exactly one minute about the types of people to consider under their given title. For example: Family might include brothers, aunts, etc. Do not use specific names, only categories or titles.
4. After one minute, each group rotates to a different paper. They have 30 seconds to see what is already on the paper, and then one minute to add more. If they can't think of any, that's okay.
5. Rotate two more times so that each group has been to each paper.
6. Bring the papers together and go over the lists. See if anyone wants to add anything.
7. Pass out the personal inventory sheets. This is the time for them to put down names. Who is part of their personal network? The brainstormed lists helps them focus on particular people.

Sample Processing Questions

- What did you notice about the brainstormed lists that were created by the group?
- What did you notice about the network that you listed for yourself?
- Why do you think it is important to identify those to whom you are connected or those to whom you can be connected? Why might a personal network be important?
- Who relies on you? Who in the world might see you as a part of their network?

Facilitation Notes

It is important for people to be aware of their own network—the people to whom they are connected or can be connected. These people are resources and potential resources. Encourage students to think about all those who affect them, to whom they would turn if they would like to talk. Who can they learn from and have good discussions with. Who can offer them help with projects? Who do they just enjoy spending time with. Remind students that relationships are reciprocal—not just about getting help or information. They are about having something to offer as well, whether it be companionship, information, or a listening ear.

YOUR PERSONAL NETWORK: VALUABLE PEOPLE IN YOUR LIFE

Family Members

People you are close to:	People who are resources to you:	Possible connection:

Community Members

People you are close to:	People who are resources to you:	Possible connection:

School People

People you are close to:	People who are resources to you:	Possible connection:

Other People

People you are close to:	People who are resources to you:	Possible connection:

Consensus: Facilitate the Pathfinder Consent activity.
Consensus is the best strategy for collaborative groups when making big decisions, but it takes practice. Many times consensus is reached quickly; other times patience and perseverance are necessary to make it work. In the end, decisions made by consensus are generally better and stronger than those made autocratically or by simple majority.

activity

PATHFINDER CONSENT

Focus: Taking turns, asking for help, consensus decision making
Materials: A large tarp with a 10 x 10 grid drawn on it with permanent marker (a grid taped to the floor can work, though it is obviously less portable), a map of the correct path for your eyes only
Time: 15 - 30 minutes
Sequence: Problem Solving
Sources: See *Adventure Education for the Classroom Community* by Frank & Panico, "Carpet Maze" in *Affordable Portables* by Cavert, "Stepping Stones" in *Games for Change* by Albin, *Journey Toward the Caring Classroom* by Frank, and "Gridlock" in *Teamwork and Teamplay* by Cain & Jolliff.

Suggested Procedure

1. Clear the desks or tables away and stand around the grid on the floor.
2. Tell the class that their task is to get everyone from one side to the other. The problem is that there is only one solid path, and it is invisible (see grid below). The other parts are quicksand.
3. To get across only one person can be on a horizontal row at any given time (this means that there can only be, at most, 10 people on the tarp at a time if it is a 10 x 10 grid). No one may skip a row or column; they may move only to a square directly to the front, front diagonal, or side (no backward moves).
4. They may not place any markers (other than their own bodies) on the tarp.
5. If someone steps on a solid point in the path, s/he receives a "thumbs up." If someone steps in a quicksand section, s/he receives a "thumbs down" and must return to the starting side.
6. **Here's the catch. No one may take a step until everyone in the group agrees to it. If someone steps without getting total group approval, he or she must return to the starting side.**

Finish

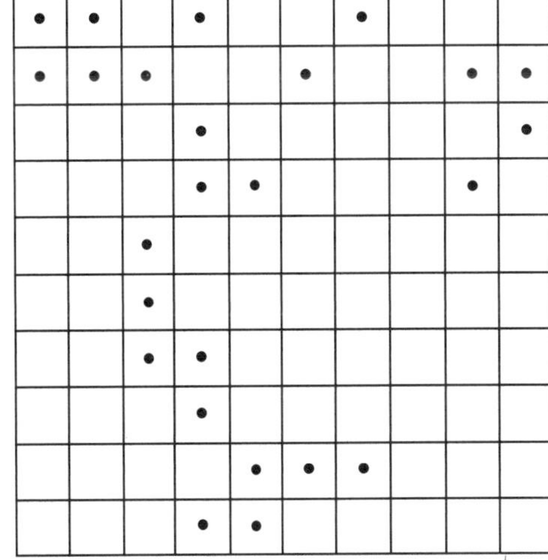

Start

Sample Processing Questions

- How easy or difficult was it for you not to step until everyone approved?
- How did you deal with this method of decision-making?
- Did you work out a system for checking in? If so, what was it? If not, how might a check-in system be helpful when trying to make consensus decisions?

Facilitation Notes

Pathfinder is a slow, yet engaging, activity that takes thought and focus. If students rush through this, it can become an exercise in frustration. The requirement to agree before taking a step can become unwieldy. This is a lesson in how consensus can take time and energy, and is not always the best decision making tool. It is, however, valuable when needing ownership from all involved.

With a larger class, try having two or more tarps, or have half the class do the activity while the other half watches (the fishbowl technique), then have them switch.

1.D^2 sequence

Journal assignment.

What made the Pathfinder Consent activity easy and/or frustrating for you? How did you like having to check in with everyone (and be checked in with) before anyone could take a step?

1.D^2 sequence

Teach consensus guidelines and Fist to Five strategies.

Discuss the meaning of consensus as a decision-making strategy. Emphasize that consensus is one way to make decisions, and is important to use when needing commitment to the tasks, goals, and vision of the group. It is appropriate to use when there is a common goal, enough time (and it does take time), and people have the skills necessary to do what is expected of them.

One definition of consensus is: **After hearing all points of view, it is the best decision for this group at this time.**[28] Acknowledge that not everyone has to agree 100%, but must feel okay about any given decision.

Guidelines for Reaching Consensus

A. Avoid arguing for your own ranking. Present your position clearly and logically. Listen to the other member's reactions and consider them carefully before you press your point.

B. Do not assume that someone must win and someone must lose when a discussion reaches a stalemate. Instead, look for acceptable alternatives for all members.

C. Do not change your mind to avoid conflict and reach agreement and harmony. Explore the reasons why decisions are made and be sure everyone accepts the solution for basically similar or complementary reasons. Yield only to positions that have logical and objective sound foundations.

D. Avoid conflict-reducing techniques such as majority vote, averages, coin flips, and bargaining. When a dissenting member finally agrees, avoid feeling s/he must be rewarded with his/her own way at a later point.

E. Differences of opinion are natural and expected. Seek them out and try to involve everyone in the decision-making process. Differences of opinion can help a group's decision by providing a wide range of information.

Fist to Five Strategy

This is a way to "vote" on a decision to see if a consensus exists.

A. State the decision that is to be made so that everyone understands.

B. Ask for a fist-to-five decision. This is where everyone puts out 0 to 5 fingers depending upon how they feel about the decision:

   5 fingers = The best idea ever!

   4 fingers = It's a really good thing to do!

   3 fingers = It's okay

   2 fingers = I can live with it

   1 finger = I won't block it

   0 fingers = Block

C. Look around to assess the quality of the decision. If no one blocks it, then a decision has been made. If, however, even one person blocks it, then more negotiating must take place before they can move on. If everyone puts out one or two fingers, then it might also behoove a group to continue working through the problem to reach a better decision.

1.D²
sequence
13

Facilitate the Batten Down the Hatches Activity.
There is also a wonderful consensus activity, called Weighty Decisions, on the Shackleton expedition that can be found on the NOVA Web site. You will find a link at: www.pbs.org/wgbh/nova/shackleton/classroom/lesson.htm

activity

Batten Down the Hatches

Focus: Consensus
Materials: A list of household items (below)
Time: 20 - 45 minutes
Sequence: Problem Solving
Sources: *Journey Toward the Caring Classroom* by Frank, "Stranded!" in *Cowstails and Cobras* by Rohnke.

Suggested Procedure

1. Give everyone a list of household items and ask them to fill it out according to the directions.
2. Then have students get into groups of 4 or 5 and give each the same list of household items.
3. The groups' task is to come to consensus on taking as many of the items as possible.
4. Remind them of the consensus guidelines and fist-to-five strategy.
5. Give them 15 to 30 minutes, depending on how much time they need and how productive their discussions and processes are.
6. Discuss the processing questions.

Sample Processing Questions

- How did you do as an individual compared with how your group did?
- Did you use the fist-to-five strategy? If so, how did it work for you?
- Why might consensus be a good decision making strategy in this case?
- When might consensus decision making not work for a group?

Facilitation Notes

Use your discretion about how much time and how much coaching the small groups need. Sometimes a little suggestion helps when a group gets bogged down. They may simply need a reminder that consensus does not mean 100% agreement, but whether or not someone can live with a decision.

Batten Down the Hatches

You are living in south Florida and have just received news that a large hurricane is heading your way. The evacuation notice has just gone out, and you have 15 minutes to gather up everything you need before having to leave. Due to limited space, you can only take 15 items with you, not including people and pets (they go automatically). The family consists of two kids, parents, and the family dog, Juno. Here is the list of supplies to choose from:

_____ Matches
_____ Five gallons of gasoline
_____ Tent with stakes and poles
_____ Case of dog food
_____ Raincoat for each person
_____ 10 pound bag of oranges
_____ Charcoal grill
_____ Charcoal
_____ Package of toilet paper
_____ Car keys
_____ Flashlight with new batteries
_____ Suitcase with a change of clothes for each person
_____ Winter coat for each person
_____ Five boxes of Pop Tarts
_____ Gallon of milk
_____ Laptop computer
_____ Weather radio
_____ Three pounds of cheese
_____ Family photo album
_____ Video game
_____ Road atlas of the United States
_____ Five gallons of water
_____ Jack knife
_____ Box of 10 candles
_____ Cell phone
_____ Emergency flares

1. As an individual, rank your top 15 items in order.
2. Once in your small groups, try and reach consensus on which 15 items to take. **You do not need to rank them in order.** If you get that far, then see if you can agree on which are the five most important ones.

1.D^2 sequence 14

Decision-making process: Introduce the STEPS[29] decision-making process.
Deciding how to decide is one thing, doing it is another. There are many processes for decision-making. The following is one that is used at Badger High School in Lake Geneva, Wisconsin. It can be used as a way to put the decision-making strategies into action.

DECISION-MAKING PROCESS

S State the problem
T Think about options (brainstorm)
E Evaluate the options
P Pick one option and try it
S Sit back and talk about the option you tried (evaluate)

Write this on the board and ask students to copy it into their journals. Have them choose a strategy for memorizing it (write it or say it over and over, make up a song or a rhythm to go with it, etc.).

In small groups, use the STEPS process to arrive at solutions to a written problem.
In small groups of 3 or 4, give each group the same dilemma (below). Using the STEPS process, have them arrive at a solution they would try. Share with the class.

Mai, Tawana, and Jason had been friends for a long time. They grew up together, went to the same schools, and spent most of their free time together. Now freshmen in high school, they were branching out more and getting involved in other things. Mai took an interest in athletics and was a star on the volleyball team. Tawana loved the visual arts and joined the art club, while Jason spent most of his time with the marching band.

About halfway through the year, they all realized that they missed each other, but there was so little time. Between their families, school, and after-school activities it seemed they would never see each other again. They decided that life is short, and it was important to figure out ways to spend time together. After all, who knew what would happen after high school was over? They could end up on opposite sides of the country or even the world!

What should they do?

Read the article on the following pages about making good decisions.

Making Decisions

One of the most important things a group can do is to decide how to decide. People make decisions all the time – from what to wear to what to do after high school. Making decisions by oneself can be difficult, making them in a group can be even more complicated.

All decisions, though, are not the same. There are four basic factors that can shape the effectiveness of a decision: 1) the type of decision being made, 2) what information is available, 3) the amount of time, and 4) the people who are making the decision. Using all four of these factors effectively can make a decision easier and more useful. If these factors are not used effectively it can interfere with the decision making process.

1) **The type of decision being made.** The bigger the problem, the more difficult the decision. A small problem, such as which day to hold a meeting, usually takes a small amount of time and energy to make. This is because the risk is low: if the wrong day is picked the worst that can happen is that no one shows up. On the other hand, if the group has to make a decision that influences whether or not they pass a class, then they should take more time and gather more information before making the decision because the stakes are higher. Big decisions also create more stress and anxiety, which, in turn, can get in the way of making a good decision. Knowing what the stakes are and staying calm are important when making decisions.
2) The second factor is **the amount and quality of information available**. The most common cause of poor decision making is probably lack of useful information. Yet in some cases, too much information can actually be a problem. Information overload causes confusion, while lack of information can cause people to make decisions that are not helpful. Be thoughtful about the information used when making decisions.
3) The third factor is **time**. Decisions always come with deadlines of some sort. The more time we have, the more information we can gather, the more we can evaluate possible solutions, and the more we can test the possibilities. A shortage of time is similar to a shortage of information because more time allows a decision maker to gather more information. In a crisis, when time is very short, all dangers seem more threatening, all decisions seem more important, and all information seems more confusing. Knowing how much time is available, and using it wisely, helps in decision making.
4) Finally, **the people who are making the decision** have an impact. The beliefs/values, skills and experience of the people involved determine the effectiveness of decisions. If the people making the decision believe in honesty, integrity, and reason they will come to a very different decision than if they are dishonest, degrading and irrational. Being under the influence of drugs or alcohol impairs judgment and makes for poor decision making. Psychologists have recognized that healthy decision-making skills are critical to the overall growth and development of young people. Such skills appear to be related to self-confidence, ability to handle stress, openness to feelings, empathy for other people's feelings, creative imagination, and ability to analyze and communicate information. Clearly, the values, skills, and resources of healthy decision makers generally contribute to effective leadership as well.

Decision Dodging, or Avoiding Making A Decision

If we put all of these factors together, we can see how they relate to the effectiveness or ineffectiveness of any decision. The best conditions for any major decision are a combination of appropriate beliefs/values, skills, and resources; moderate stress; enough information; and ample time. Even when all the factors line up, there are some ways that groups avoid making difficult decisions: 1) procrastinating, 2) passing the buck, or 3) wanting to keep things the same.

1) **Procrastinating** is simply putting off the decision until some future time—perhaps forever. It's a very common way to avoid difficult decisions, because it can be used whenever a decision maker feels no pressure to act right away. The trouble with procrastination is that problems often get worse when we ignore them.
2) **Passing the buck** is simply trying to get someone else to make the decision. Decision makers may pass the buck if they are unwilling to make a decision but unable to postpone it. Leaders are often the victims of buck-passing, since people often feel justified in passing their problems up to their bosses or leaders. Effective leaders, however, encourage followers to take responsibility for decisions whenever possible. They also accept ultimate responsibility for group decisions whenever necessary. As Harry Truman said about his duty as President of the United States: "The buck stops here."
3) The final form of avoidance is **wanting to keep things the same**, which generally tries to pretend the problem doesn't exist. People will do this when they lose hope of making a better decision but are unable to procrastinate or pass the buck.

Try Not to Panic

Besides avoiding a decision, one more pitfall comes in the form of panic. Panic can strike when decision makers feel that the problem is very serious but there is not enough time. Panic causes people to seek information frantically, but it interferes with the ability to interpret information or use it wisely. People trying to escape from a burning building may be examples of panic, as they run from one escape route to another or pick the first choice without adequate thought. In large groups, people experiencing panic can become their own worst enemies, as they make each other more afraid and stampede each other in a desperate attempt to solve a problem or avoid danger. Decisions made during panic are like decisions made under the influence of alcohol or drugs—they may be worse than no decision at all.

Making Good Decisions

Procrastination, buck-passing, wanting things to stay the same, and panic all usually result in bad decisions because they short-circuit the decision making process. Unless decision makers are just plain lucky, significant decisions work out well only if the decision making process has been careful and thorough. Specifically, decision makers must diagnose the problem and the need for change, thoroughly investigate a wide range of possible solutions, survey all the goals and values at stake in the decision, compare all the costs and benefits of each possible solution, consider all the doubts and possible regrets raised by the final choice, and recognize the need for commitment to the chosen course of action.

We'll close with a few suggestions for effective decision making in leadership roles.

- **Prevention:** Remember that the only way to avoid stressful decisions is to prevent problems ahead of time. Encourage group members to be on the lookout for problems, to discuss them openly, and to keep learning so new problems don't sneak up on them.
- **Make Decisions Together:** Get as many ideas as possible. The more involved people are in the decision, the more they want to help make it happen. Be open to all ideas, even if they don't seem useful right away. An idea that seems strange at first may be just what is needed later.
- **Avoid Groupthink:** All the advantages of group decision making can be cancelled out if group members are too eager to please each other and conform to group standards. Encourage respectful disagreement and constructive criticism, and invite outside experts to consult on group decisions whenever possible.
- **Use a Decision Making Process:** Use a format like the STEPS process to help organize the way the group makes a decision. It can help keep everyone on track.

Not all decisions work out. So if you find yourself stuck with a bad decision, don't waste time feeling foolish or ashamed—just go out and fix it by making a better decision.

1.D² *sequence* 17

In small groups, return to the Shackleton and *Into Thin Air* articles to find examples of decision making (see p. 34 for the Shackelton article, and p. 38 for *Into Thin Air*).
Have small groups peruse the articles and find at least 5 examples discussed in the previous, decision making article of good and/or bad decision making. Have each group share their findings with the class.

1.D² *sequence*

Journal assignment.
What did you learn about how decisions are made in your life? Do you want to change how you make decisions? If so, how? If not, how is it working for you?

Lesson 1.D³ Overview: Conflict Resolution Strategies

Focus: Encourage diversity of opinion
Encourage participation
Resolve conflicts

Step	Activity	Time	Assessment
1	**Recognizing conflict:** Facilitate the Group Role Play activity.	10 - 15 min	
2	Using the processing questions from the activity, discuss how people handled the simulated conflict.	10 - 15 min	
3	Facilitate the Conflict Web activity.	15 - 30 min	
4	Share Glasser's five basic needs.	10 - 20 min	
5	Facilitate the Conflicting Needs activity.	20 - 30 min	assessment
6	Journal assignment.	10 - 15 min	assessment
7	Present the conflict escalator.	10 - 20 min	
8	**Conflict escalation and dealing with anger:** Read "A True Story of Road Rage" out loud.	10 - 15 min	
9	Draw a conflict escalator on the board and chart the road rage article.	10 - 20 min	
10	Introduce the concept of anger mountain.	5 - 10 min	
11	Ask students to complete a conflict escalator using a conflict they have had, and discuss in a small group.	15 - 20 min	assessment
12	Ask students to discuss their conflict escalators with a partner and then in a small group using the provided questions and tasks.	15 - 30 min	assessment
13	Journal assignment.	10 - 15 min	self assessment
14	**Seeking win-win solutions:** Facilitate the Thumb Wrestling activity.	5 - 10 min	
15	Introduce the concept of win-win solutions.	10 - 20 min	
16	Read the *Zax* story by Dr. Seuss.	10 - 15 min	
17	Journal assignment.	10 - 15 min	assessment
18	Introduce STEPS as a conflict resolution process.	5 - 10 min	
19	Practice STEPS as a conflict resolution process.	15 - 30 min	
20	Create before-and-after skits to show how a conflict can be handled poorly and well using the skills.	15 - 30 min	assessment

Time frame: 3.5 - 7 hours

Lesson 1.D³ Sequence: Use a Variety of Skills and Tools for Collaboration: Conflict Resolution Strategies

Recognizing Conflict: Facilitate the Group Role Play activity.

activity

Group Role Play

Focus: Recognizing conflict
Materials: None
Time: 10 - 15 minutes
Sequence: Deinhibitizer/Trust
Sources: Laurie Frank learned this activity from Sandy Whittall.

Suggested Procedure

1. Create two lines facing each other. Each person should be facing his/her partner.
2. Have each pair decide who will play him/herself and who will be "the sibling."
3. Present the following scenario that involves a minor conflict between two people.
   The night before, you asked your sibling if you could use the bathroom first in the morning because you had to leave early for a school trip. S/he agreed. When you woke up, s/he was in the bathroom, and taking his/her time, too. You pound on the door and say....."
4. Give them a few minutes to act out the scenario all at the same time.

Sample Processing Questions

- What were some thoughts/feelings you had about yourself, your sibling, the situation?
- What started the conflict?
- Did you say or do anything that made the situation better or worse?

Facilitation Notes

Although rare, there are times when someone says something that can be taken personally and bring up a real conflict. Make sure to present this activity in a fun and light way. Emphasize that this is simply a game and not a real conflict. People should take care not to say or do anything that really hurts someone's feelings.

Using the processing questions from the activity, discuss how people handled the simulated conflict.

- What were the similarities and differences between the pairs and the way they handled it?
- Did it bother some more than others?
- Why do they think people handled it the way they did?

1.D[3]
sequence
3

Facilitate the Conflict Web activity.

activity

Conflict Web

Focus: Exploring the nature of conflict
Materials: A ball of yarn, chalkboard/whiteboard and markers, or butcher paper and markers
Time: 15 - 30 minutes
Sequence: Deinhibitizer/Trust
Sources: "Support Web" in *Building Assets Together* by Roehlkepartain, "Spider Web" in *Tribes: A New Way of Learning and Being Together* by Gibbs.

Suggested Procedure

1. Sit in a circle, either in chairs or on the floor.
2. Write the word "Conflict" in the middle of the board and circle it.
3. Pose the following thoughts and questions:
   - *Think of a time when you have been involved in a conflict. It could have been with a sibling, like in our role-play. It could have been with parents or other family members, friends, strangers, adults, or peers. It could have been with one person or a group of people.*
   - *What was the conflict about? What caused it? Was it resolved, did it simply go away, or is it still a conflict?*
   - *Now think of your thoughts and feelings about the conflict. Think of one word or phrase that describes some of your thoughts and feelings.*
4. Tell the group they are going to make a web, and hand the ball of yarn a participant.
5. The person with the yarn says their word or phrase and then, holding onto the yarn, throws the ball to someone else in the circle.
6. The person who catches the yarn then states her word or phrase, throws the ball while holding onto the ball, then holds on to the yarn and throws the ball to someone else.
7. This continues until everyone has had the yarn and stated their word or phrase.
8. As each person says their word or phrase, write it on the board, connecting it to the word "conflict," making a visual "web" of the words as they connect to each other.
9. Remind people that they have the right to pass. They can keep hold of the yarn, say "pass," and throw the yarn ball to someone else. In this way, they are still part of the web.
10. Ask students to look at the web they created from their words, and to look at the web that they created with the yarn. Ask them to consider the following statement:
    "Conflict can be like this web. If we are not careful, we can get caught in it. Conflict, in itself, is neither good bad. It all depends on how the conflict is dealt with."
11. Explain that they are going to unwind the web by going in reverse order. Ask them to think about a time when a conflict worked out for the good or was left unresolved or worked out poorly.
12. As they pass the yarn back through their reverse order, have them wind it up and, if they would like, to share about a time when a conflict worked out well, was unresolved, or did not work well. They have the right to pass.

Sample Processing Questions

- Is there anything you notice about the words that were written on the board?
- Do you usually see conflict as a good thing or a bad thing? What causes you to see it that way?

- Are conflicts always useful? Are they always harmful? Why or why not?
- What makes a conflict a good thing or a bad thing?

Facilitation Notes

Make sure students know that they always have a right to pass. The issue of conflict may be too threatening for some students, and they need to know that they will not be required to share information that they are not ready to share.

It is also important to monitor what people are sharing. If it becomes too personal, you may need to step in and redirect the discussion. For example, if a student begins to talk about being hit by a parent, you can say something such as, "You might want to talk to me about that later, is there another conflict that you would like to share?" It is then necessary to check in with that student. Mandatory reporting guidelines will dictate that you take action along with the school social worker.

Share Glasser's five basic needs (from Choice Theory).

William Glasser asserts that human beings have five basic needs that drive behavior.[30] A list of these needs and characteristics is on the following page. Everyone has all of these needs and, depending upon whether or not these needs are being met, they may seek to meet one or more of them. Glasser talks about how these needs are like being thirsty. We don't have a choice about it, and we must satisfy the need eventually. Therefore, if we are able to find water soon, then the thirst need goes away for the time being, and we can focus on other, more pressing, needs.

It follows, then, that if someone is feeling the need for freedom (maybe she is grounded at home, or has been working so much that she doesn't have much free time), she may chafe at rules that are imposed at school, or by someone who has a high need for power at the moment and is trying to control how people work together on a project.

Glasser's Five Basic Needs (from Choice Theory)

Need	Characteristic (if this need is high)
Survival	• Takes few risks • Makes lots of future plans • Wants to keep things as they are (status quo) • Relies on common sense
Belonging	• Seeks connection with others • Enjoys meeting new people and the company of others • Wants to be included in activities • Is concerned with other people's welfare
Freedom	• Doesn't like rules • Takes risks • Wants time to be alone • Doesn't like crowds and congestion
Power	• Seeks respect of others • Needs to be taken seriously and listened to • Seeks control of many situations • Avoids being "wrong"
Fun	• Is interested in everything • Frequently laughs and smiles • Delights in new knowledge • Actively pursues recreation

Facilitate the Conflicting Needs activity.

activity

Conflicting Needs

Focus: Identifying possible conflict situations
Materials: Needs descriptions
Time: 20 - 30 minutes
Sequence: Problem Solving

Suggested Procedure

1. Divide the class into five groups and ask each group to sit together.
2. Present the following scenario:

   You have been chosen to be on the student council. The first gathering will be a weekend-long retreat so that the new council can choose some goals for the year and elect its officers. When you get to the retreat center, you are assigned to a room with three other people. There is much work to do, but there is also time for people to socialize and get to know each other.
3. Give each group one of the following descriptions of a person going to the retreat.
4. Ask each group to decide which of the five basic needs this person may have (hint: like being thirsty, look for things that are missing in that person's life at the moment), and write down what he or she might do to get those needs met while on the retreat. Remind them to think about both meeting time and social time. How might their person react in both scenarios?
5. Ask each group to share with the class.
6. As a class, determine if any of these people might come into conflict with others who have competing needs. Where and when might a conflict arise? (See processing questions.)
7. Role play some of the situations and discuss ways the situations could be handled to resolve the conflicts.

Person A: You are a senior, and this is your fourth year on student council. Getting into a good college is important to you. You have been studying very hard and have just completed your SATs and ACTs. It seems like you have not had a night out or a day off forever. Your parents think you take yourself too seriously and keep telling you to lighten up. You are well liked in school and have a very large social group with whom you hang out at school every day. Last year you were student council vice president and in charge of organizing the prom. It was a huge success, and you hope it will help you get elected as president this year.

Person B: You are a freshman, and this is your first year on student council. High school has been a little better than middle school, where your grades weren't so great and you spent quite a bit of time in the office. Even though things are better this year, your parents don't trust you. You are not allowed to participate in extra-curricular activities because your parents keep you at home to study. Getting your homework done is the most important thing to them, but not to you. At the request of your parents, you had to give up your favorite subject, art, so you could take an extra study hall.

Person C: You are a junior, and this is your second year on student council. You come from a close family of 12 brothers and sisters. You live in an apartment that is too small for your family and you share a room with five of your siblings. Your parents work very hard to make sure you have what you need, but it's always a stretch to make ends meet, especially toward the end of the month. You feel hungry most of the time. You secretly think about giving up extra-curricular activities like student council to get a job and help out the family.

Person D: You are a senior, and this is your first year on student council. In fact, you are surprised that people voted for you because you don't have many friends. You have been a loner during high school. People steer clear of you because of your political views. You even decided to be a non-conformist in your dress after people made fun of you during middle school. Even so, you love to learn and enjoy your classes very much, even though you don't participate in the discussions. During the last year you have shown a real aptitude for computer graphics and have become the person who designs the school newspaper. You plan on going to college for a degree in graphic design.

Person E: You are a sophomore, and this is your second year on student council. You are a wonderful abstract thinker, but most people don't take you seriously; usually because your ideas are complicated and you have a difficult time communicating them. Last year on student council you were silenced over and over, until you finally stopped offering ideas. You're still not sure why you ran for student council this year, but are hoping things will be different. You have a knack for juggling and magic, which fascinates people, and which you have found can be a bridge between your complicated thinking and making friends.

Sample Processing Questions

- How did the lives of each of these people affect how they reacted at the retreat?
- How might their needs cause conflicts with others who have very different needs at the moment?
- If any of these students can get one of their needs met, how might that change the way they react at the retreat?
- What might be a good way to help resolve some of these conflicts?

Facilitation Notes

There are no right or wrong answers in this activity. Many of the student descriptions indicate more than one unmet need. The main emphasis here is for students to think about the idea of basic needs and how they might influence the way people do things. These needs sometimes compete with each other and cause conflicts. For example, someone with a high need for freedom may come into conflict with someone who has a high need for belonging, as he or she may see that person as "clingy."

The other emphasis is that the importance of any given need changes depending on what is going on in one's life. Sometimes we can get needs met easily, while at other times it seems impossible to address a need. Like thirst, it will not go away, and the urge to satisfy the need will remain until something changes.

6

Journal assignment.

What is the difference between needing something and wanting something? How can one's wants cause conflict? How can one's needs cause conflict?

Conflict Resolution Strategies[31] curriculum by Carol Miller Lieber introduces some of the basics of conflict resolution. Recognizing conflict and resolving conflict are two different things. It is necessary to gain the skills to resolve conflict and practice them. We recommended that you purchase this resource (see bibliography).

1.D³ sequence 7

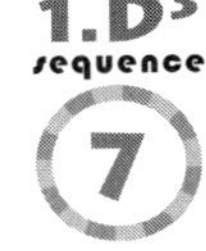

Present the Conflict Escalator.

Educators for Social Responsibility articulates the concept of the Conflict Escalator. The idea is that conflict does not automatically get worse, and we have choices about how a conflict progresses. Many times, our choices cause conflicts to "escalate" or get worse. If people hop onto the conflict escalator, then a simple act of dropping a pencil can end with people getting suspended:

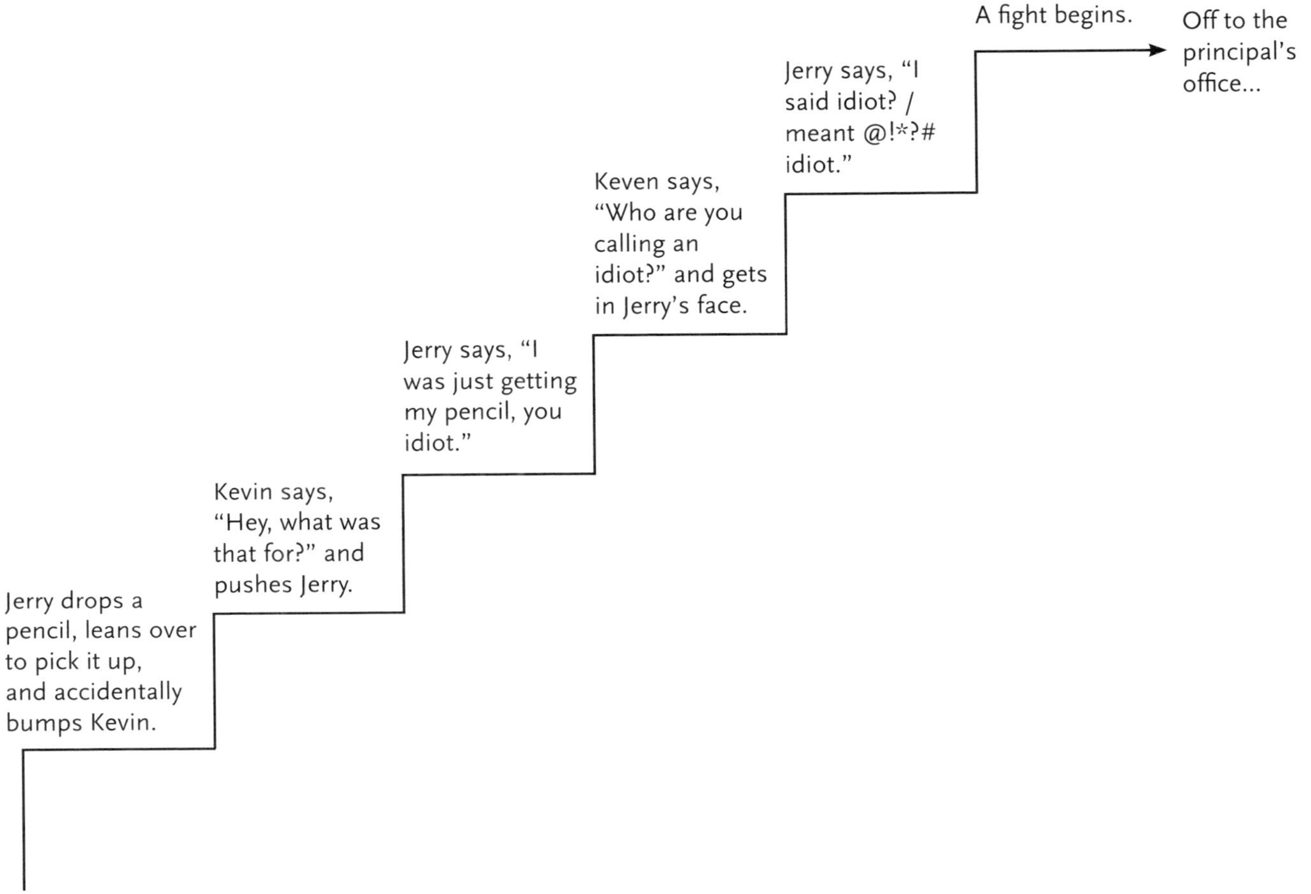
Jerry drops a pencil, leans over to pick it up, and accidentally bumps Kevin.
Kevin says, "Hey, what was that for?" and pushes Jerry.
Jerry says, "I was just getting my pencil, you idiot."
Keven says, "Who are you calling an idiot?" and gets in Jerry's face.
Jerry says, "I said idiot? / meant @!*?# idiot."
A fight begins.
Off to the principal's office...

Conflict escalation and dealing with anger: Read the following true story about road rage out loud.

Road Rage

One day I was driving on the highway and one of my tires blew. It was scary for a moment because I started fishtailing, but was able to get it under control and move off to the side of the road. After I stopped shaking, I started to think about the next steps. That's when I noticed that there was a car behind me. A big guy was getting out of the car, and it was obvious that he was irate. When he got to my window, I opened it just a bit. He yelled, "What do you think you were doing out there?! Where did you learn to drive?"

Although scared and irritated, I tried to remain calm, "I am so sorry. My tire blew out."

Instead of offering to help, the man went off on me, "Your lousy car shouldn't be on the road. Why don't you take better care of it," he screamed.

"Really," I said, "I'm sorry. I'll pay for any damage to your car." I just wanted to get out of there, but this guy wouldn't quit.

He shouted even louder, "Dumb jerks like you give all drivers a bad name."

I felt my face get hot, and before I knew it I was out of the car and in his face, "You have some nerve! My tire blew and I almost got killed. It should have flown off and hit your car."

He got into my face then and even gave me a little shove, shouting, "People like you shouldn't be allowed out of your house, never mind behind the wheel of a car!"

"You bald-headed idiot," I shouted back. "Get out of my face so I can fix my car!"

In response, he grabbed me and shoved me up against my car. I kicked and hit anything in my way.

That's when the red lights came up behind us. When the officer approached, she asked, "What's going on here?"

Both of us were bleeding and hurt, but we were able to point at each other and say, "You started it!"

We were both ticketed for disorderly conduct.

1.D³ sequence

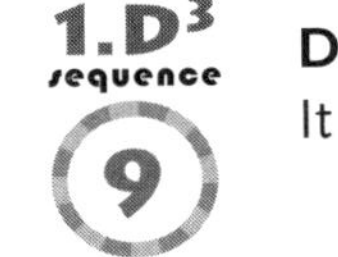

Draw a conflict escalator on the board and chart the Road Rage article.
It might look something like this:

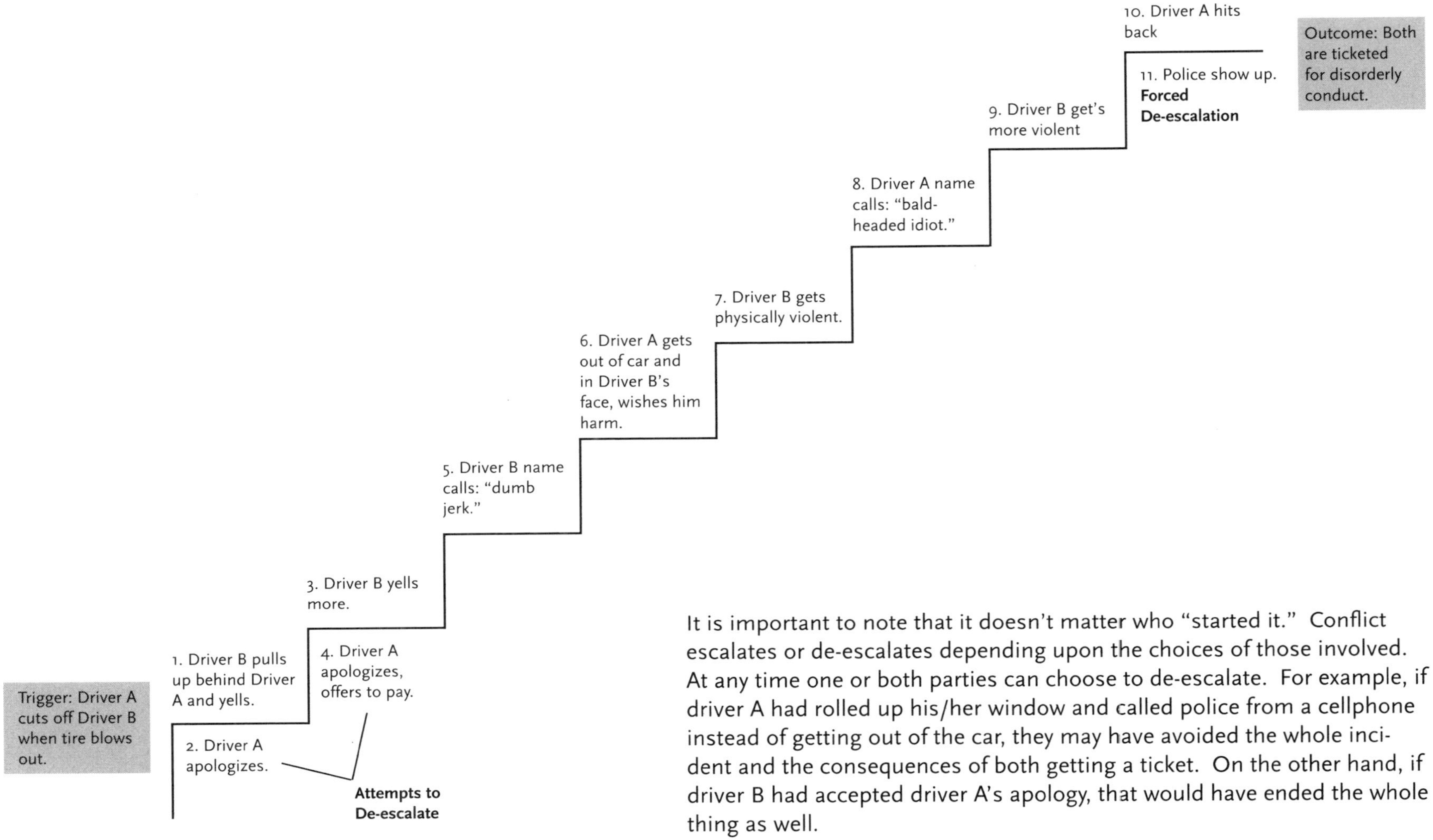

Trigger: Driver A cuts off Driver B when tire blows out.

1. Driver B pulls up behind Driver A and yells.
2. Driver A apologizes.
3. Driver B yells more.
4. Driver A apologizes, offers to pay.

Attempts to De-escalate

5. Driver B name calls: "dumb jerk."
6. Driver A gets out of car and in Driver B's face, wishes him harm.
7. Driver B gets physically violent.
8. Driver A name calls: "bald-headed idiot."
9. Driver B get's more violent
10. Driver A hits back
11. Police show up. **Forced De-escalation**

Outcome: Both are ticketed for disorderly conduct.

It is important to note that it doesn't matter who "started it." Conflict escalates or de-escalates depending upon the choices of those involved. At any time one or both parties can choose to de-escalate. For example, if driver A had rolled up his/her window and called police from a cellphone instead of getting out of the car, they may have avoided the whole incident and the consequences of both getting a ticket. On the other hand, if driver B had accepted driver A's apology, that would have ended the whole thing as well.

Conflict Escalator[33]

Conflict: How was it/could it have been resolved?

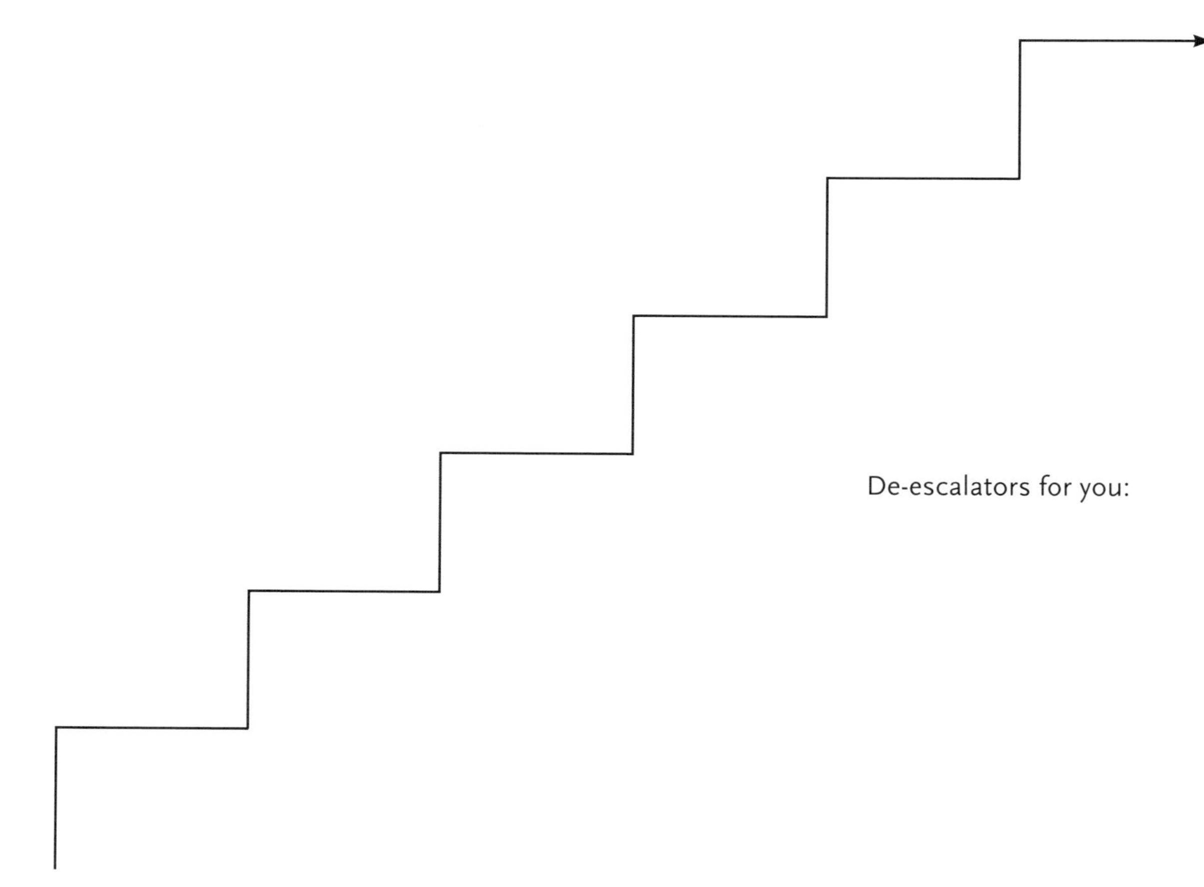

De-escalators for you:

1.D³ sequence 10

Introduce the concept of Anger Mountain.

The main concept with Anger Mountain is that we have about seven or eight seconds once a trigger is hit before we lose control. This is a physiological response to anger. In order to stay off the anger escalator, one must be aware of what is going on and de-escalate quickly. An article in the Contact Syracuse newsletter of the Central New York School Violence Prevention Network describes what it means to "climb the anger mountain":

> Once the hot button sets us off, there are only seven seconds before adrenaline kicks in, leading to an explosion. As the physiological response takes over, thinking stops and the primitive brain rules. Once the explosion is over, we can calm down and think again. It is important for adults and youth to know what their hot buttons are and to practice responses that interrupt the chemical reaction and explosion. Deep breathing, self talk, thought stopping, and time out are some methods that students can practice the next time they feel themselves becoming angry (Contact Syracuse, 2000).[32]

Using the escalator chart, ask students to think of a conflict they have had with at least one other person and chart how it escalated.

If students are having a difficult time reporting a personal conflict, allow them to make one up. The intent here is to assess that they understand how conflict can escalate.

Ask students to discuss their conflict escalators with a partner and then in a small group using the provided questions and tasks.

1. With a partner, discuss the conflict you charted.
2. If the conflict was resolved, how was it resolved?
3. If the conflict was not resolved, how could it have been resolved?
4. What are some of your "triggers"? What causes you to get angry pretty fast? For example, some people become angry when someone lies, while another person doesn't mind that as much as name-calling.
5. When you become angry or upset, what are some de-escalators that help you calm down so that you can deal with the conflict? **Write each one on a sticky note.** Example: Count to 5.
6. As a small group of 4 to 6, post your de-escalators (sticky notes) on a sheet of paper.
7. Put the ones that are similar in groups together.
8. Label each group of notes. For example, if you have "listen to music," "play my clarinet," and "singing" in a group, you could label it as "Music."

Have each small group share their de-escalation categories with the class. Combine and post them on the wall for future reference.

Journal assignment.

Which of the de-escalators works best (or might work best) for you? How will you know to use the de-escalators the next time one of your "triggers" is pushed? Record your reaction in your journal.

1.D3 sequence 14

Seeking win-win solutions: Facilitate the Thumb Wrestling activity.

activity

Thumb Wrestling

Focus: Comparing competition and cooperation
Materials: None
Time: 5 - 10 minutes
Sequence: Ice Breaker/Deinhibitizer
Sources: Laurie Frank first learned this activity from Craig Dobkin during a Play for Peace event.

Suggested Procedure

1. Teach or review the fine art of thumb wrestling: Have partners reach out like they are shaking hands, curling the fingers inward, and nesting the fingers together. In this way both partners' thumbs should be closest to the ceiling/sky, not closest to the floor/ground. The object is to get your partner's thumb pinned against his hand using only your thumb.
2. Have everyone pair up, and give them 30 to 60 seconds to thumb wrestle to the highest number of points.
3. After the first round ask everyone what their score was.
4. If someone has a very high score, have them show how they got it.
5. If not, ask how it might be possible to maximize the scores (it is possible to achieve scores of 20 or even 50).
6. Work toward the idea of cooperation. When they work together by taking turns putting each other's thumbs down, both people can get very high scores.

Sample Processing Questions

- What was your common goal?
- What caused people to compete with each other rather than cooperate?
- What happened when people chose to cooperate?
- How did that create a win-win situation, rather than a win-lose situation?

Facilitation Notes

Even people who are cooperative by nature tend to fall into the competitive mode when presented with this activity. It is a familiar game that many have played competitively in the past, and they continue to do it in the same way even when the goal has changed. Other times it has to do with the competitive culture of the society. This activity helps people think about when competition does not work and look for other ways to arrive at mutually beneficial solutions.

Sometimes there are one or two pairs that figure out how to get more points by cooperating. Other times it is necessary to help students change their way of thinking about it.

$1.D^3$ sequence 15

Introduce the concept of win-win solutions.

Scenarios for resolving conflict can play out in a variety of ways:

> **Lose-Lose:** We both want the last piece of cake and are fighting over it. Someone in authority comes by, sees the argument, and takes the cake for herself. Neither of us get cake.
>
> **Win-Lose:** Person A got there first so he gets the cake (or whoever is the strongest, most insistent, etc., gets the cake).
>
> **Lose-Win:** We are fighting over the cake and one of us gives up and allows the other person to have the whole thing.
>
> **Win-Win:** We share the cake, or we make a cake together and share it, or...

Win-win solutions mean that when people disagree, they are able to find a solution that is beneficial for all parties. In this scenario, the parties in conflict search for ways that each side can get its needs and goals met.

$1.D^3$ sequence

Read the *Zax* story by Dr. Seuss.[34]

High schoolers can enjoy a good children's book. This one does a wonderful job of showing a lose-lose situation. After reading it, have students offer suggestions for turning it into a win-win situation.

$1.D^3$ sequence

Journal assignment.

Bring back the scenario from the beginning of the unit and ask students to brainstorm win-win solutions:

> *The night before, you asked your sibling if you could use the bathroom first in the morning because you had to leave early for a school trip. S/he agreed. When you woke up, s/he was in the bathroom, and taking his/her time, too. You pound on the door and say...."*

$1.D^3$ sequence

Introduce STEPS as a conflict resolution process.

This process was used as a decision-making strategy, but can also be used to help people work through conflict. When a conflict arises and people have de-escalated and are ready to talk it out, the STEPS process can be used as a model.

S State the problem

T Think about options (brainstorm)

E Evaluate the options

P Pick one option and try it

S Sit back and talk about the option you tried (evaluate)

Practice STEPS as a conflict resolution process.

Ask everyone to write down a conflict they have had or witnessed. Have them fold it up and throw it into a container. Mix them up and have someone draw one from the container. Read it to the class and then go through the STEPS process to come up with solutions.

This can also be done in small groups.

In small groups, create before-and-after skits to show how a conflict can be handled poorly and well using the skills.

Have the class break into small groups. Remind them about conflict escalating, dealing with anger, win-win solutions and the STEPS process. Ask them to create a skit with two acts—one where the conflict does not get resolved or resolved well, and the other where the conflict is resolved well.

If you would like to give your class more structure, you can use some of the conflict suggestions from step 19.

Lesson 1.D4 Overview: Goal Setting Strategies

Focus:
Foster a climate of continuous learning
Set measurable outcomes
Balance group needs with individual needs
Practice reflection

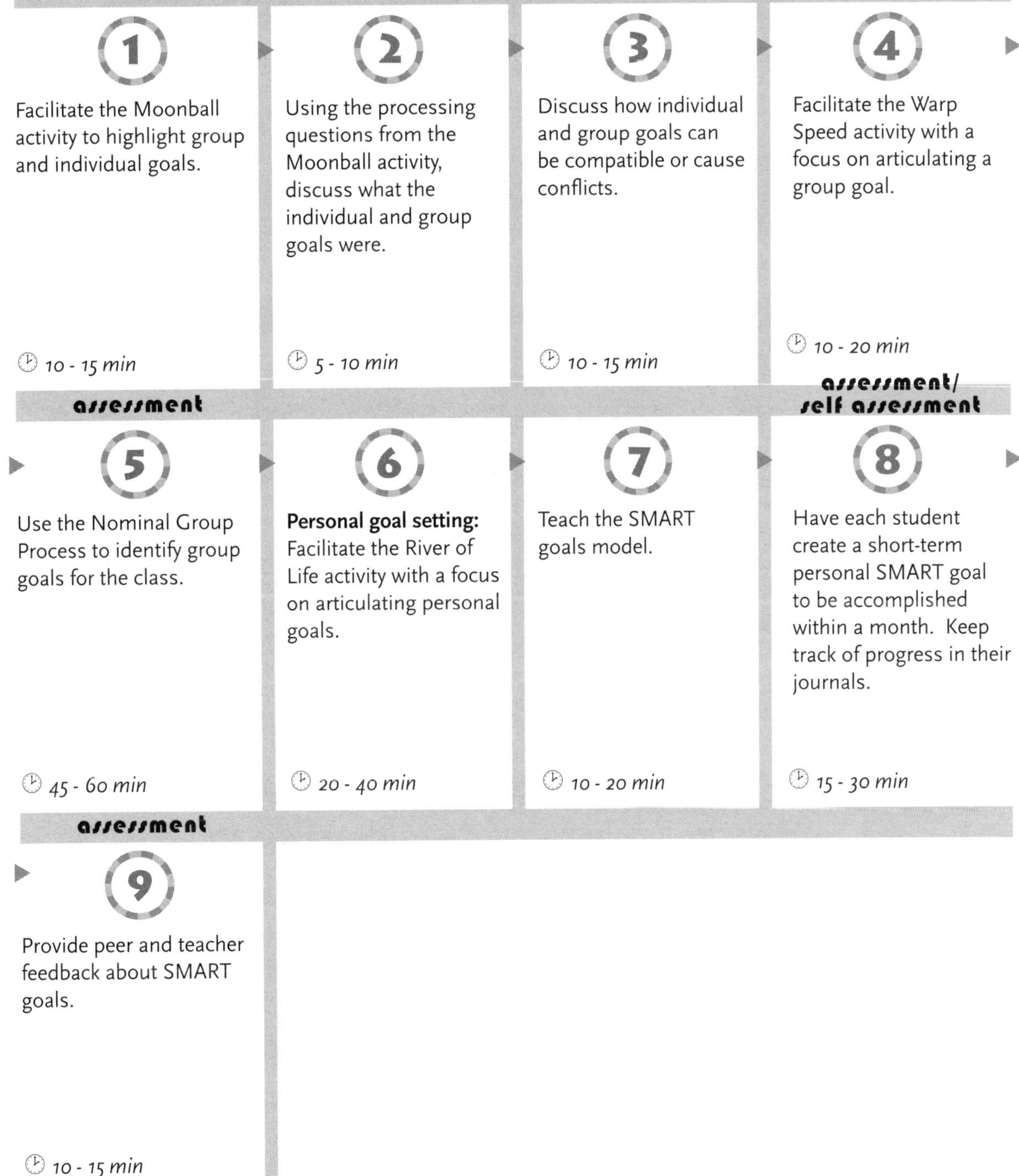

LESSON 1.D4 SEQUENCE: USE A VARIETY OF SKILLS AND TOOLS FOR COLLABORATION: GOAL SETTING STRATEGIES

Facilitate the Moonball activity to highlight group and individual goals.

activity

MOONBALL

Focus: Cooperation, problem solving, goal setting, conflict resolution, leadership, organization, decision making
Materials: A beach ball
Time: 10 - 15 minutes
Sequence: Problem Solving
Sources: See *Adventure Education for the Classroom Community* by Frank & Panico, *Affordable Portables* by Cavert, *Journey Toward the Caring Classroom* by Frank, *Quicksilver* by Rohnke & Butler, and *Silver Bullets* by Rohnke.

Suggested Procedure

1. Clear the desks or tables away and stand in a circle.
2. Tell the students that the object is to bounce the beach ball in the air. Each time the ball is hit, it counts as a point, and the ball cannot be hit by the same person twice in a row. If the ball stops or touches the ground, they start over. Then throw the ball in the air for them to start.
3. After they have tried some rounds, adjusted their strategies and had some success with this, ask them to set a goal for themselves.
4. Continue until they achieve the goal. Then set a new one.
5. If they try over and over but still cannot meet their goal, ask if they would like to adjust their goal.

Sample Processing Questions

- What goals did you set for yourselves? Did it help to set goals? Why or why not?
- Before you set the group goal, what goal were you personally striving for? Were these individual goals compatible?
- Did you feel you were included in this activity? Why or why not?
- What were some strategies you used to achieve your goals? How did these strategies change over time?

Facilitation Notes

The rules to Moonball are vague by design. This provides an opportunity for the class to discuss a variety of strategies and interpretations. Generally they will begin with the "chaos method." Sometimes this actually works. Replicating the results is another issue altogether. Finally, they will begin to arrive at an agreed-upon scheme to become more consistent.

A common influence issue that must be addressed is one of ownership—making sure everyone has an opportunity to be involved, rather than a few people hitting the ball while others stand around and watch. Encourage your class to set some goals. They might find that a goal gives them a place to go, rather than hitting a beach ball for no other apparent reason.

If you have a low ceiling or lots of furniture in the room, an interesting dynamic can occur. Participants may blame the low ceiling or the amount of furniture for their lack of success. If this happens, take a look at how trying to control things that are truly out of their control may not be the best strategy. See if they can focus on what they can control and see what happens. Generally they can meet with more success. A valuable lesson!

1.D4 sequence 2

Using the processing questions from the Moonball activity, discuss what the individual and group goals were.

1.D4 sequence 3

Discuss how individual and group goals can be compatible or cause conflicts.
Introduce the concept of hidden agendas—the conflict between a personal need or goal and the group's needs and goals. Many times hidden agendas arise when an individual's personal goals or needs are not compatible with the group goals and needs at the time. For example:

> A person in authority (boss, board president, etc.) creates the agenda for a meeting, does all the talking in the meeting, and generally makes decisions without input from others in the group. In this case, the group members' need for power (Glasser) may not be met or certain members may be excited about ideas or expertise (personal goals) they do not get to share. If there is no way for them to bring up their concerns, then it is likely that the actions the group takes will be met with apathy or even sabotage. The group members' agendas will then be hidden. If, on the other hand, there is a way to bring up the issues in a small-group conference with the person in authority, a listening session with the whole group, or a team-building day facilitated by an outside person, then the hidden agendas can possibly be exposed and can be dealt with in the open.

1.D4 sequence

Facilitate the Warp Speed activity with a focus on articulating a group goal.

activity

WARP SPEED

Focus: Common sense, cooperation, goal setting
Materials: One soft throwable object (fleece ball, foam ball, wadded up paper), stop watch
Time: 10 - 20 minutes
Sequence: Problem Solving
Sources: See *Adventure Education for the Classroom Community* by Frank & Panico, *Adventures in Peacemaking* by Kreidler & Furlong, *Changing the Message* by Albin, *Diversity in Action* by Chappelle & Bigman, "Eggspediency" in *Quicksilver* by Rohnke & Butler, and *Journey Toward the Caring Classroom* by Frank.

Suggested Procedure

1. Stand in a circle.
2. Ask everyone to raise one hand to show that they have not had the object yet.
3. Call someone's name and throw the object to her. She puts her hand down to show she's had the object and calls the name of someone whose hand is up and throws him the object. This continues until everyone has had the object and it is returned to you. Thus, a pattern is set.
4. Figure out who has the next birthday in the class. Give that person the object.

5. Tell the group that this is a timed activity, and they must send the object to the same person they threw it to before. It must begin and end with the person who starts it (in this case, the person with the next birthday). Try this and get a baseline time.
6. Now tell the group that the rules are this: The object must touch everyone in that same order, and it must begin and end with the same person.
7. Ask them to set a group goal.
8. Then give them time to discuss strategies.
9. Try the activity multiple times to arrive at a mutual solution.

Sample Processing Questions

- Describe your trial and error approach. What steps did you take to arrive at your final solution?
- How did you evaluate each attempt? Did you then use that lesson to arrive at your next solution?
- How did your group goal affect the way you solved this problem?
- How did you decide which idea to try?
- How did your solution change each time? Were you willing to learn from each attempt?

Facilitation Notes

It is fair to say that Warp Speed has an almost limitless number of solutions, which is what makes the activity so popular. It can also be accomplished in a relatively short amount of time.

Many groups decide to stand next to each other, rather than stay in their original configuration. Other groups stay put and use different ways to get the object around without moving themselves. One class on an overnight weekend decided their group goal would get the object around as slowly as possible. They figured out how long each person should keep the object, then sent it around during the rest of the three-day experience. Students set their alarms to go off in the middle of the night just to pass "Mr. Corn" to the next person in the sequence.

Use the Nominal Group Process to identify group goals for the class.

The nominal group process is a way to help groups set goals and begin to develop action plans. Although it takes some advance preparation, it helps to balance participation so that everyone has an equal voice in the creation of group goals. It is also helpful in that it provides an avenue for proactive problem solving rather than the reactive stance of simply complaining about the problem.

Ask students what they would like to see as group goals for this class. Have them think and reflect on their relationships, ground rules, accomplishments, and disappointments. Are people getting their work done, are they satisfied with how the class is going? Ask them to go back over their journals to see if that triggers any ideas.

Once they have had the time to reflect, then start the nominal group process. These goals do not need to be life-changing, but something that they are willing to work toward. If people are having a difficult time generating ideas individually, spend some time brainstorming in small-groups.

THE NOMINAL GROUP PROCESS by Boris Frank

Nominal Group Process is an excellent formal method of establishing a priority for organizational goals. To be successful, the steps should be carried out in the following order. Do not anticipate the next step, as that might alter the results.

1. Ask the group to spend approximately five minutes silently thinking about and writing down their top three organizational goals. They should be stated in as few words as possible, for example, "Draft marketing plan," "Explore partnerships."
2. Go around the group, one at a time, and ask each person in order to list **one** of their proposed goals. There should be no discussion at this point, other than questions of clarification. Each recommendation should be succinctly listed by the leader on a flip chart and **lettered** in consecutive order (A, B, C, etc.).
3. Once everyone has made a recommendation, go around the group again, taking only **one** recommendation from each person. If all of a person's ideas are on the board, he/she should pass. Do **not** add a goal after the listing process has begun. The whole point is to begin focusing on the most important goals. If anyone would like to suggest minor alterations or add to another person's recommendation, this may be allowed with the permission of the person who made the original suggestion. If the person does not agree, the new recommendation should be listed separately.
4. Continue until all ideas are on the board.
5. Once all recommendations have been listed, the group is allowed to discuss them. They may ask questions for clarification, lobby for a particular idea, or suggest that some of the ideas be combined. **Be very careful about combining, however.** Combining usually results in very broad categories, which become universal truths rather than relatively specific goals. Resist combining ideas unless they really are so similar that they demand to be combined. All parties must agree that the combination is acceptable.
6. Once discussion has been concluded—and you will probably have to conclude it yourself for the sake of time—ask each participant to write down the letter of his or her top **five** recommendations—**BUT NOT IN ANY PRIORITY ORDER**. It is important they not put them in priority order yet.
7. Ask the group to put the number 5 next to the most important recommendation. **They should not do anything else at this time—trying to anticipate the next step could skew results.** The procedure is important and the group should be asked to follow it step-by-step.

   Next ask participants to select the lowest priority of the four remaining on their list and place the number 1 next to it.

   Of the remaining three choices, place a 4 next to the **most** important.

   Of the remaining two choices, place a 2 next to the **less** important of the two.

   And finally, place a 3 next to the remaining recommendation.
8. Go around the room one-by-one and ask participants to read off their selections, giving recommendation point scores for each on their list. Keep a tally on your flip chart. After everyone has reported their selections, total the score for each recommendation.
9. In analyzing results, look for both total score (that will determine order) and the number of participants voting for a particular recommendation. Three people may give a 5 to an item for a total score of 15. On the other hand, eight or nine people could vote for an item and the score could total substantially less than 15. Weigh results carefully, realizing that items need not be discarded simply because the score is low.
10. Now is the time to combine items for the purposes of implementation and putting together specific action objectives.
11. Determine process objectives for each goal—"Who will do what next to implement the goal?"

Personal Goal Setting: Facilitate the River of Life activity with a focus on articulating personal goals.

activity

The River of Life

Focus: Personal goal setting, risk taking, physical/emotional trust, trustworthiness
Materials: Two long ropes for boundaries (clothesline works fine), lots of "stuff" to put inside the boundaries (wadded up pieces of paper, stuffed animals, shoes, etc.), a sticky note and writing utensil for each person, eye coverings (optional)
Time: 20 - 40 minutes
Sequence: Trust
Sources: See *Adventure Education for the Classroom Community* by Frank & Panico, "Conflict Field" in *Adventures in Peacemaking* by Kreidler & Furlong, "Minefield" in *Affordable Portables* by Cavert, "Life's a Journey" in *Changing the Message* by Albin, "Minefield" in *Diversity in Action* by Chappelle & Bigman, "Passages" in *Games for Change* by Dodds & Prosser-Dodds, "Challenge Field" in *Games for Teachers* by Cavert & Frank, *Journey Toward the Caring Classroom* by Frank, "Pitfall, 3-D Minefield & Minefield in a Circle" in *Quicksilver* by Rohnke & Butler, and "Minefield" in *Silver Bullets* by Rohnke.

Suggested Procedure

1. Clear out the middle of the room, use a large area like a gym, or go outside.
2. Place the two ropes on the ground parallel to each other, about 10 to 15 feet apart. Make them wavy to simulate a river.
3. Distribute the "stuff" randomly inside the "river" boundaries.
4. Ask each student to think about a personal behavioral goal—something they wish to work toward in this class. For example, maybe someone is very quiet in a large group and wishes to be more vocal. That person may set a goal to contribute to a group discussion at least once per day. Another student may have trouble getting her homework in on time. Her goal may be to finish all her homework. Have them write their goal on a sticky note.
5. Divide the class into pairs and stand around the "river of life."
6. Explain that the river of life is full of accomplishments and frustrations. One way to navigate through the river of life is to set goals for oneself. This helps to provide direction. Many times, though, there are obstacles to achieving goals—some are external, some self-imposed. For example, if a goal is to get one's homework in on time, some obstacles might be: procrastinating by choosing to do other things (like playing games or watching TV), other responsibilities at home, or leaving it at school by mistake. The obstacles are represented by the stuff that is strewn about.
7. Have each pair choose someone to be the goal-getter first. The other is the guide. The goal-getter places his/her goal in the river of life. The object is to get to the goal, pick it up, and get to the other side, touching as few obstacles as possible.
8. The goal-getter dons an eye-covering or closes her/his eyes. The guide then verbally directs her/him to the goal. If an obstacle is touched by the goal-getter, then s/he tells the guide about an obstacle that s/he might encounter when trying to achieve the particular goal.
9. After reaching the other side, the partners switch roles.

Sample Processing Questions

- What did your guide do to help you achieve your goal?

- What are some resources you have available to help you achieve your goal? These can involve things or people.
- What are some obstacles you are likely to encounter when working toward your goal? What can you do to deal with the obstacles so they do not get in the way of achieving your goal?
- How will you know when you have achieved your goal? How will you keep track? Is there anyone who can help you track your goal?

Facilitation Notes

Make sure you have done some activities that require eye coverings before this one so that the students are already used to the concept of being guided by someone else. Also, remind goal-getters to keep their hands in the bumpers-up position.

This activity can get loud and confusing for the students. Some teachers prefer to have the guides stay on the outside of the boundaries, while others permit the guides to accompany the goal-getters on the inside. This decision should be based upon the maturity and experience of the students. Also, the amount of stuff on the inside will be determined by the ability of the class. The more items, the more difficult the activity becomes.

1.D4 sequence 7

Teach the SMART goals model.

Goal setting really is an art. SMART goals help to focus a goal to make it action-oriented and concrete. Without such clarity, a goal can easily become overwhelming.

SMART GOALS:

S **Specific:** No "either/or" statements. Say exactly what you want to achieve. Set one goal; you can always amend it later if it does not fit any more.

M **Measurable:** Time and quantity – When? How much? You know when you have achieved the goal.

A **Achievable/Attainable:** Realistic – Stretch yourself, but don't frustrate yourself. Perfection isn't necessary.

R **Relevant/Results Oriented:** Must make sense in this context, and have an end in mind.

T **Trackable/Time-Bound:** Monitor progress and know when goals are accomplished.

Examples of SMART Goal:

Behavioral/Personal

Vague Goal	SMART Goal
I will stop being messy.	I will clean my bedroom every Saturday for the next three months.
I'll read more.	I will read at least three books per month between July 1 and December 31.
I want to save money toward college.	I will put 10% of my paycheck into savings each month for the next three years.

Academic/Professional

Vague Goal	SMART Goal
Get my homework done.	I will complete and hand in 95% of my assignments for my classes this year.
Get a job.	I will apply for three jobs every week for the next month.
Do well in school.	I will get at least 80% on my geometry assignments this semester.

1.D4 sequence 8

Have each student create a short-term personal SMART goal to be accomplished within a month. Keep track of progress in their journals.

1.D4 sequence 9

Provide peer and teacher feedback about SMART goals.
Using the form provided, have students share their SMART goal with at least two other people, as well as yourself, for feedback.

Academic/Professional
Peer & Teacher Evaluation

Goal writer:______________________ Evaluator:______________________

Criteria	Evaluation*	Comments
Is this goal specific?		
Is this goal measurable?		
Is this goal achievable/ attainable?		
Is this goal relevant/ results oriented?		
Is this goal trackable/ time-bound?		

Other comments:

* Evaluation key: + = Excellent, o = Okay, - = Needs work.

Lesson 1.E Overview: Apply Collaborative Skills

Focus: Encourage participation
Practice reflection

	1	2	3	4
Label				assessment/ self assessment
Activity	Review the Unit 1 skills and concepts.	Small group projects to practice collaborative skills and concepts.	Before planning the project, have the class create an assessment rubric for how they would like to be evaluated on their presentations.	Small group presentations to the class.
Time	*15 - 30 min*	*1 - 3 hr*	*10 - 30 min*	*varies: 20 - 25 min per group (includes evaluation)*

	5	6	7
Label	self assessment	self assessment	assessment
Activity	Journal questions.	Students chart portfolio completion using the chart provided.	Students and teacher evaluate portfolio.
Time	*10 - 15 min*	*20 - 40 min*	*varies*

Time Frame: 3.5 - 7 Hours, plus portfolio time

LESSON 1.E SEQUENCE: APPLY COLLABORATIVE SKILLS

1.E sequence 1

Review Unit 1 skills and concepts.
Ascertain if there are any questions or gaps.

The Collaborative Process Skills/Themes	Notes
Developing Group Cohesion • Group building activities • Shackleton's Voyage • Into Thin Air • Qualities of productive and non-productive teams • Social contract	
Diversity • Multiple intelligences • Being an ally • Group identities • Honoring diversity	
Collaboration • Definitions of cooperation and collaboration • Similarities and differences between cooperation and collaboration	
Skills/Tools for Collaboration • Communication: Active listening, non-verbal cues • Decision-Making: Types of decisions, brainstorming, consensus, STEPS process • Conflict Resolution: Recognizing conflict, conflict escalation and anger, win-win solutions, STEPS process • Goal Setting: Group goals, nominal group process, personal goals, SMART goals	

1.E sequence 2

Small group projects to practice collaboration skills and concepts.

Task: In groups of 3 to 5, students need to plan and teach a 15-minute lesson.

Guidelines:

- Create groups either randomly or according to interest.
- Have each small group plan a 15-minute presentation for the entire class.
- Have small groups brainstorm possibilities and come to consensus on one thing they will teach to the class.
- In the process, have them create a small group goal and a personal SMART goal for their presentation. They are to use at least two different media: one oral and the other either visual (PowerPoint, posters, etc.) or interactive (song, role play, activity).

1.E sequence 3

Before planning the project, have the class create an assessment rubric for how they would like to be evaluated on their presentations.

What criteria are important to show quality work? Have them think about the quality of the presentation, how the group worked together, etc. It may look something like this:

Criteria	1	2	3
Participation	One or two people did all the work.	Everyone did something, but the workload was uneven.	We shared responsibility.
Collaboration	Everyone worked alone.	We shared some information.	We created the presentation together.
Goals	We didn't try to meet our goals.	We met some part of our goals.	We made a good effort to meet our goals and met most/all of them.
Quality	We did everything at the last minute and threw it together.	We put it together in pieces and it worked.	We really created a fine presentation.
Etc.			

Presentation Evaluation Rubric

Criteria	1	2	3

1.E sequence 4

Present lessons to the class.

Students should rate the lessons using the agreed-upon criteria (rubric) along with self-evaluation using the same rubric.

Journal assignment.

Look at your group and personal goals for this project. How did you do? Did you achieve your goals? Explain.

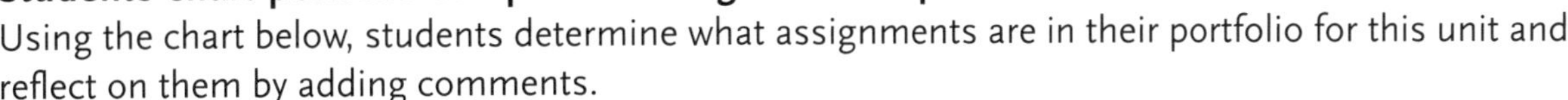

Students chart portfolio completion using the chart provided.

Using the chart below, students determine what assignments are in their portfolio for this unit and reflect on them by adding comments.

The purpose of this exercise is to help students remember what they have done and reflect on their learning. This is a list of assignments and assessments throughout the unit. There can certainly be more items in the portfolio than are listed here, especially if a student wishes to use it to keep track of his or her work.

Portfolio Completion Chart

Item (Lesson)	Comments
Ice breaker, deinhibitizer, trust, and problem solving activities (1.A)	
Journal: Describe your experience of getting to know the people in this group. What were your feelings when you came into this class on the first day? What are they now? (1.A)	
Multiple Intelligence Inventory. (1.B)	
Journal: Choose one or more: How has your uniqueness been helpful or hurtful to you in school and in your life? How have you been an ally, and/or so you wish to be an ally with others? Who do you look down upon and how would your life be different if you decided to become their ally? What have you been picked on for and what would you like your allies to say or do about this? (1.B)	
Research someone who has gained or lost (or both) because of his or her group identity. (1.B)	
Journal: Using your base team definitions of collaboration rewrite the definition to create a personal definition of collaboration. (1.C)	
Create a rubric with some of the collaboration concepts and rate yourself on collaborative skills. (1.C)	
Rate your active listening skills. (1.D[1])	

PORTFOLIO COMPLETION CHART (CONT.)

Item (Lesson)	Comments
Creativity: Four Measures of Style brainstorming activity. (1.D²)	
Personal Network Brainstorming activity. (1.D²)	
Journal: What made the Pathfinder Consent activity easy and/or frustrating for you? How did you like having to check in with everyone (and be checked in with) before anyone could take a step? (1.D²)	
Journal: What did you learn about how decisions are made in your life? Do you want to change how you make decisions? If so, how? If not, how is it working for you? (1.D²)	
Journal: What is the difference between needing something and wanting something? How can one's wants cause conflict? How can one's needs cause conflict? (1.D³)	
Conflict escalator: De-escalators. (1.D³)	
Journal: Which of the de-escalators works best (or might work best) for you? How will you know to use the de-escalators the next time one of your "triggers" is pushed? Record your reaction in your journal. (1.D³)	
Journal: Bring back the scenario from the beginning of the unit and ask students to brainstorm win-win solutions: *The night before you asked your sibling if you could use the bathroom first in the morning because you had to leave early for a school trip. S/he agreed. When you woke up, she was in the bathroom and taking his/her time, too. You pound on the door and say. . ..* (1.D³)	

PORTFOLIO COMPLETION CHART (CONT.)

Item (Lesson)	Comments
Personal SMART Goal. (1.D4)	
Small group class presentations. (1.E)	
Journal: Look at your group and personal goals for this project. How did you do? Did you achieve your goals? Explain. (1.E)	
Other:	
Other:	

1.E sequence 7

Students and teacher evaluate portfolio.

Evaluation will depend on the purpose of the class and the culture in the school. One way to evaluate portfolios is to have each student convene a group of 3 peers. The student then chooses a sampling of his or her work to share with this review committee. Their task is to look over the work and meet with the student to ask questions and provide feedback about the quality of the work. Before doing this you may wish to create some ground rules about how feedback will be offered. Using your established social contract is an option, as well.

The same sampling of work can then be evaluated by the student him/herself, and the teacher. If necessary, grades can be suggested by each of the parties, and a grade can be assessed for the portfolio.

[1] Check out NOVA's Web site, and explore 360-degree panoramic images of South Georgia and its waters, and journey to places virtually unchanged in the decades since Shackleton's expedition, on PBS.org. Educators can order this or any other NOVA program: call WGBH Boston Video at 1-800-255-9424. NOVA's *Shackleton's Voyage of Endurance* Web site includes lesson ideas and commentary. http://www.pbs.org/wgbh/nova/shackleton/. *Shackleton's Voyage of Endurance* DVD can also be purchased through Amazon.com.

[2] *Shackleton's Voyage of Endurance*. (2002, March 26). Retrieved March 19, 2004, from http://www.pbs.org/wgbh/nova/transcripts/2906_shacklet.html

[3] Krakauer, J. (1996). *Into Thin Air* [Electronic version]. Outside Magazine, September.

[4] Krakauer, J. (1997). *Into Thin Air: A personal account of the Mt. Everest Disaster.* New York: Random House.

[5] A hearty "thank you" to Marilyn for her willingness to share her skills and materials for use in this curriculum. You can find out more about Marilyn at www.marilynlevin.com

[6] Thanks to Marilyn Levin for these ground rules when dealing with sensitive issues.

[7] This survey was developed by, and is used with permission of, Dell Coats-Erwin © 1999. The Multiple Intelligence name and concept was developed by Dr. Howard Gardner of Harvard University.

[8] Thanks to Carol Carlin for the Human Bar Graph strategy.

[9] Adapted from: Gardner, H. (1999). *Intelligence reframed: Multiple intelligences for the 21st century.* New York: BasicBooks. Gay, G. (n.d.), *Eight styles of learning.* Retrieved April 12, 2007, from Learning Disabilities resource community, http://www.ldrc.ca/projects/miinventory/miinventory.php?eightstyles=1

[10] Thanks to Marilyn Levin for help with these questions.

[11] Levin, M. (2007). Unpublished manuscript.

[12] Chappelle, S. & Bigman, L. (1998). *Diversity in Action: Using adventure activities to explore issues of diversity with middle school and high school age youth.* Beverly, MA: Project Adventure, Inc., p. 127. Used with Permission.

[13, 14, 15, 16, 17, 18] Levin, M. (2007). Unpublished manuscript.

[19] *Cambridge Advanced Learner's Dictionary* (2007). Retrieved April 17, 2007 from http://dictionary.cambridge.org/define.asp?key=17007&dict=CALD

[20] Olsen, K. & Pearson, S. (2000). *Character begins at home.* Kent, WA: Susan Kovalik and Associates.

[21] Goleman, D. (1997). *Emotional intelligence: Why it can matter more than IQ.* New York: Bantam Books.

[22] Leader to Leader Institute (2006). Formerly the Drucker Foundation. Retrieved, April 17, 2007 from http://www.leadertoleader.org/knowledgecenter/collab_challenge/challenge/appendixa.html

[23] Schrage, M. (1990) *Shared minds.* New York: Random House.

[24] Panitz, T. (1996). *A definition of collaborative vs. cooperative learning.* Retrieved April 17, 2007 from http://www.city.londonmet.ac.uk/deliberations/collab.learning/panitz2.html

[25] Adapted from Givens, D. (2006). *The Nonverbal Dictionary of Gestures, Signs & Body Language Cues.* [electronic version]. Center for Nonverbal Studies. Retrieved April 18, 2007 from http://members.aol.com/nonverbal2/diction1.htm. Used with Permission.

[26] Adapted from *Creativity: Four Measures of Style.* Badger High School Leadership Curriculum (n.d). Used with permission.

[27] Adapted from *Your Personal Network.* Badger High School Leadership Curriculum (n.d.)

[28] Thanks to Carol Carlin for this definition.

[29] Adapted from Badger High School Leadership Curriculum (n.d.)

[30] This theory has some similarities to Maslow's Hierarchy of Needs, which submits that people have a series of innate needs that range from most to least powerful in this order: physiological, safety, love/belonging, esteem, and self-actualization.

[31] *Conflict Resolution in the High School* by Carol Miller Lieber presents a comprehensive approach to conflict resolution with high-school students. Please see the bibliography to find out where to obtain it.

[32] Anger Management: One Size Does NOT Fit All. (May, 2004). Retrieved April 18, 2007 from www.contactsyracuse.org

[33] Thanks to Educators for Social Responsibility for the concept.

[34] The Zax is found in the book *Sneetches* by Dr. Seuss.

"We are bound together by the task which lies before us."

—Martin Luther King

UNIT 2 outline

What is Leadership?

To intentionally grow one's collaborative leadership skills, it is helpful to construct a paradigm or model of leadership. This mental picture allows young people to deliberately obtain skills, seek out mentors, and pay attention to subtleties they may otherwise miss. Becoming a skillful collaborative leader is a never-ending journey, and making one's way with eyes wide open is not only more efficient, but it can make the trip that much more stimulating, exciting, and fun.

By the end of this unit, students will create their own vision of an ideal leader. In preparation for this task, students undertake an exploration of leadership. It begins with a return to nurturing the relationships in the group. In Unit 1 there was a focus on group cohesion, where students were participants in the process. Unit 2 puts students in the role of facilitator, attending to the sequence of activities that help create a sense of community.

Students will devise their own personal definition of leadership. This definition will be revisited throughout the course as more information and experiences are assimilated. It is a working definition that will change over time and serve as a metaphor for how one's own proficiency can evolve. With definitions in hand, students grapple with the dichotomy of whether leaders are born or made, and eventually debate the matter. This leads into some traditional notions of leadership, Situational Leadership®, as well as an investigation into the role of risk taking and vision when taking on a leadership role. All of these explorations prepare students for their final task of the unit: constructing a vision of an ideal leader.

A. Enduring Understandings

- There are many theories about leadership.
- Learning about leadership is a lifelong journey that begins with the question: "What is leadership?"
- We all have role models whose leadership qualities we admire.
- Leadership involves risk taking to varying degrees.

B. Essential Questions

- What are some traditional and current theories of leadership?
- What role does risk taking play in being a leader?
- Can anyone lead, or is it something reserved for a few?
- Who are recognized leaders in history, and what are their leadership qualities?
- Who is a leader that I admire, and what are his/her leadership qualities?
- How does one learn to be a leader?

C. Key Knowledge and Skills

Participants will know:

- Traditional theories of leadership.

- The theory of Situational Leadership®.
- How leadership can involve taking risks.
- The nature of a vision and mission statement.

Participants will be able to:

- Recognize and lead icebreaker activities with a small group.
- Identify one (or more) leader they consider to be a role model and distinguish the qualities they admire about that person.
- Identify one (or more) recognized leader in history and distinguish the qualities that made him/her an effective leader.
- Identify their "ideal" leader.

D. Unit Activities/Performance Tasks: (Experiential Lessons)

- Creating community.
- Compose a personal definition of leadership.
- Compare and contrast the notions that leaders are born and/or made.
- Explore different views of leadership.
- Analyze the roles of risk taking and long-term vision when leading.
- Construct a vision of an ideal leader.

E. Varied Classroom Assessments and Rubrics

- Jumble Activity.
- Facilitation rubric to evaluate how each small group facilitated an activity.
- Journal assignments.
- Write about a person they consider to be a role model as a leader, and the traits they admire.
- Write a SMART goal about developing a life-skill.
- Video guide question sheet about aristocracy and democracy.
- Skits showing the difference between an aristocratic and democratic scenario.
- Debate about whether leaders are born or made.
- Attempt to reach consensus on whether leaders are born or made.
- Write a personal definition of leadership.
- Revisit and rewrite working definition of leadership.
- Create examples of when different styles of leadership are useful.
- Research one person and his/her leadership style(s) as they relate to traditional, Situational Leadership®.
- Create a collage depicting their chosen leader's styles.
- Create a personal vision and mission statement.
- Evaluation of visual representation of ideal leader using created rubric.
- Portfolio evaluation.

F. Student Self-Assessment

- Facilitation rubric to evaluate how each small group facilitated an activity.
- Journal assignments.
- Chart progress of life-skill development through a SMART goal.
- Revisit and rewrite definition of leader.
- Evaluation of visual representation of ideal leader using created rubric.
- Portfolio evaluation.

G. Peer Feedback

- Facilitation rubric to evaluate how each small group facilitated an activity.
- Debate about whether leaders are born or made.
- Gallery walk of collages about a chosen leader and his/her styles.
- Evaluation of visual representation of ideal leader using created rubric.
- Portfolio evaluation.

H. Logistics

- Time Frame: 21 to 34+ hours
  The time frames presented here are intended only as a guide. Each teacher has his or her own style, and every class has its own personality and needs. The amount of time invested in each part of the lesson is dependent upon choices the teacher makes about which activities to present, what is assigned for homework or done in class, richness of the discussion, etc.
- Materials to gather (other than copies of handouts): Please see each activity for intended use.
  - Soft throwable objects (e.g., stuffed animals, wadded up pieces of paper) 1/person
  - 3- to 12-foot lengths of rope (1/person)
  - Eye coverings, aka: blindfolds (optional)
  - 8- x 10- foot tarp
  - Hula hoop or short rope
  - Stop watch
  - Journals (1/person)
  - One or more videos: *Leadership through identification: Kundun, The Madness of King George, The Queen, The Story of Rosa Parks, Norma Rae, Gandhi*
  - A raw egg for each group of 4 to 5 (a couple extra to allow for breakage), 25 straws
  - 3 feet of masking tape for each group, garbage bag, large paper and markers
  - Magazines with pictures for collage-making
  - Ropes or tape for boundary markers (3 12-foot ropes from above)

UNIT 2 relevant standards

What is Leadership?

McRel Standards and Benchmarks*:

BEHAVIORAL STUDIES

Standard 1: Understands that group and cultural influences contribute to human development, identity, and behavior.

1. Understands that cultural beliefs strongly influence the values and behavior of the people who grow up in the culture, often without their being fully aware of it, and that people have different responses to these influences.
2. Understands that social distinctions are a part of every culture, but they take many different forms (e.g., rigid classes based solely on parentage, gradations based on the acquisition of skill, wealth, and/or education).
3. Understands that people often take differences (e.g., in speech, dress, behavior, physical features) to be signs of social class.
5. Understands that the difficulty of moving from one social class to another varies greatly with time, place, and economic circumstances.
1. Understands that heredity, culture, and personal experience interact in shaping human behavior, and that the relative importance of these influences is not clear in most circumstances.
7. Understands that family, gender, ethnicity, nationality, institutional affiliations, socioeconomic status, and other group and cultural influences contribute to the shaping of a person's identity.

Standard 2: Understands various meanings of social group, general implications of group membership, and different ways that groups function.

1. Understands that while a group may act, hold beliefs, and/or present itself as a cohesive whole, individual members may hold widely varying beliefs, so the behavior of a group may not be predictable from an understanding of each of its members.
2. Understands that social organizations may serve business, political, or social purposes beyond those for which they officially exist, including unstated ones such as excluding certain categories of people from activities.
4. Understands that groups have patterns for preserving and transmitting culture even as they adapt to environmental and/or social change.
5. Understands that social groups may have patterns of behavior, values, beliefs, and attitudes that can help or hinder cross-cultural understanding.

Standard 3: Understands that interactions among learning, inheritance, and physical development affect human behavior.

3. Understands that expectations, moods, and prior experiences of human beings can affect how they interpret new perceptions or ideas.

---

4. Understands that people might ignore evidence that challenges their beliefs and more readily accept evidence that supports them.

LANGUAGE ARTS: Writing

Standard 1: Uses the general skills and strategies of the writing process.

2. Evaluates own and others' writing (e.g., accumulates a body of written work to determine strengths and weaknesses as a writer, makes suggestions to improve writing, responds productively to reviews of own work).
9. Writes persuasive compositions that address problems/solutions or causes/effects (e.g., articulates a position through a thesis statement; anticipates and addresses counter arguments; backs up assertions using specific rhetorical devices [appeals to logic, appeals to emotion, uses personal anecdotes]; develops arguments using a variety of methods such as examples and details, commonly accepted beliefs, expert opinion, cause-and-effect reasoning, comparison-contrast reasoning).
11. Writes reflective compositions (e.g., uses personal experience as a basis for reflection on some aspect of life, draws abstract comparisons between specific incidents and abstract concepts, maintains a balance between describing incidents and relating them to more general abstract ideas that illustrate personal beliefs, moves from specific examples to generalizations about life).

Standard 2: Uses the stylistic and rhetorical aspects of writing.

2. Uses paragraph form in writing (e.g., arranges paragraphs into a logical progression, uses clincher or closing sentences).
6. Organizes ideas to achieve cohesion in writing.

Standard 4: Gathers and uses information for research purposes.

3. Uses a variety of print and electronic sources to gather information for research topics (e.g., news sources such as magazines, radio, television, newspapers; government publications; microfiche; telephone information services; databases; field studies; speeches; technical documents; periodicals; Internet).
7. Scans a passage to determine whether it contains relevant information.

LANGUAGE ARTS: Listening and Speaking

Standard 8: Uses listening and speaking strategies for different purposes.

1. Uses criteria to evaluate own and others' effectiveness in group discussions and formal presentations (e.g., accuracy, relevance, and organization of information; clarity of delivery; relationships among purpose, audience, and content; types of arguments used; effectiveness of own contributions).
4. Asks questions as a way to broaden and enrich classroom discussions.
3. Uses a variety of strategies to enhance listening comprehension (e.g., focuses attention on message, monitors message for clarity and understanding, asks relevant questions, provides verbal and nonverbal feedback, notes cues such as change of pace or particular words that indicate a new point is about to be made; uses abbreviation system to record information quickly; selects and organizes essential information).
4. Adjusts message wording and delivery to particular audiences and for particular purposes (e.g., to defend a position, to entertain, to inform, to persuade).

5. Makes formal presentations to the class (e.g., includes definitions for clarity; supports main ideas using anecdotes, examples, statistics, analogies, and other evidence; uses visual aids or technology, such as transparencies, slides, electronic media; cites information sources).
7. Uses a variety of verbal and nonverbal techniques for presentations (e.g., modulation of voice; varied inflection; tempo; enunciation; physical gestures; rhetorical questions; word choice, including figurative language, standard English, informal usage, technical language) and demonstrates poise and self-control while presenting.
8. Responds to questions and feedback about own presentations (e.g., clarifies and defends ideas, expands on a topic, uses logical arguments, modifies organization, evaluates effectiveness, sets goals for future presentations).
9. Understands influences on language use (e.g., political beliefs, positions of social power, culture).

LANGUAGE ARTS: Viewing

1. Uses a range of strategies to interpret visual media (e.g., draws conclusions, makes generalizations, synthesizes materials viewed, refers to images or information in visual media to support point of view, deconstructs media to determine the main idea).
2. Uses a variety of criteria (e.g., clarity, accuracy, effectiveness, bias, relevance of facts) to evaluate informational media (e.g., web sites, documentaries, news programs).

LIFE SKILLS: Self-Regulation

Standard 1: Sets and manages goals.

6. Sets explicit long-term goals and shorter range subgoals.
9. Sets routine goals for improving daily life.
11. Displays a sense of personal direction and purpose.

Standard 2: Performs self-appraisal.

13. Identifies personal motivational patterns, personality characteristics, and style.
14. Identifies desired future accomplishments and preferred lifestyle.
15. Identifies significant life experiences, including key accomplishments, successes, and peak experiences.

Standard 5: Maintains a healthy self-concept.

4. Demonstrates appropriate responses to criticisms (e.g., avoids overreacting, takes criticism in a dispassionate manner).
9. Analyzes criticisms to determine their accuracy and identifies useful lessons learned.

Standard 6: Restrains impulsivity.

1. Keeps responses open as long as possible.
2. Remains passive while assessing situation.
3. Suspends judgment.

LIFE SKILLS: Thinking and Reasoning

Standard 1: Understands and applies the basic principles of presenting an argument.

5. Uses a variety of strategies to construct an argument (e.g., facts, anecdotes, case studies, quotations, logical reasoning, tables, charts, graphs).
6. Evaluates the overall effectiveness of complex arguments.

7. Evaluates an argument objectively by considering all sides of an issue (e.g., using past experience, data, logical analysis.
8. Knows that a good argument anticipates and attempts to answer objections before they are posed.
9. Develops a clear and coherent thesis and conclusion for an argument.

LIFE SKILLS: WORKING WITH OTHERS

Standard 1: Contributes to the overall effort of a group.

1. Knows the behaviors and skills that contribute to team effectiveness.
2 Works cooperatively within a group to complete tasks, achieve goals, and solve problems.
2. Demonstrates respect for others' rights, feelings, and points of view in a group.
6. Identifies causes of conflict in a group and works cooperatively with others to deal with conflict though negotiation, compromise, and consensus.
7. Helps the group establish goals, taking personal responsibility for accomplishing such goals.
8. Evaluates the overall progress of a group toward a goal.
9. Contributes to the development of a supportive climate in groups.

Standard 4: Displays effective interpersonal communication skills.

1. Demonstrates appropriate behaviors for relating well with others (e.g., empathy, caring, respect, helping, friendliness, politeness).
4. Exhibits positive character traits towards others, including honesty, fairness, dependability, and integrity.
7. Responds to speaker appropriately (e.g., does not react to a speaker's inflammatory deliverance, maintains objectivity, reacts to ideas rather than to the person presenting the ideas).
10. Uses emotions appropriately in personal dialogues.

Standard 5. Demonstrates leadership skills.

2. Knows the qualities of good leaders and followers.
3. Knows a variety of leadership strategies, and knows which strategies to implement in specific situations.

LESSON 2.A OVERVIEW: CREATING COMMUNITY

FOCUS: Create a safe environment for collaboration
Encourage participation
Foster a climate of continuous learning
Choose the appropriate facilitation technique

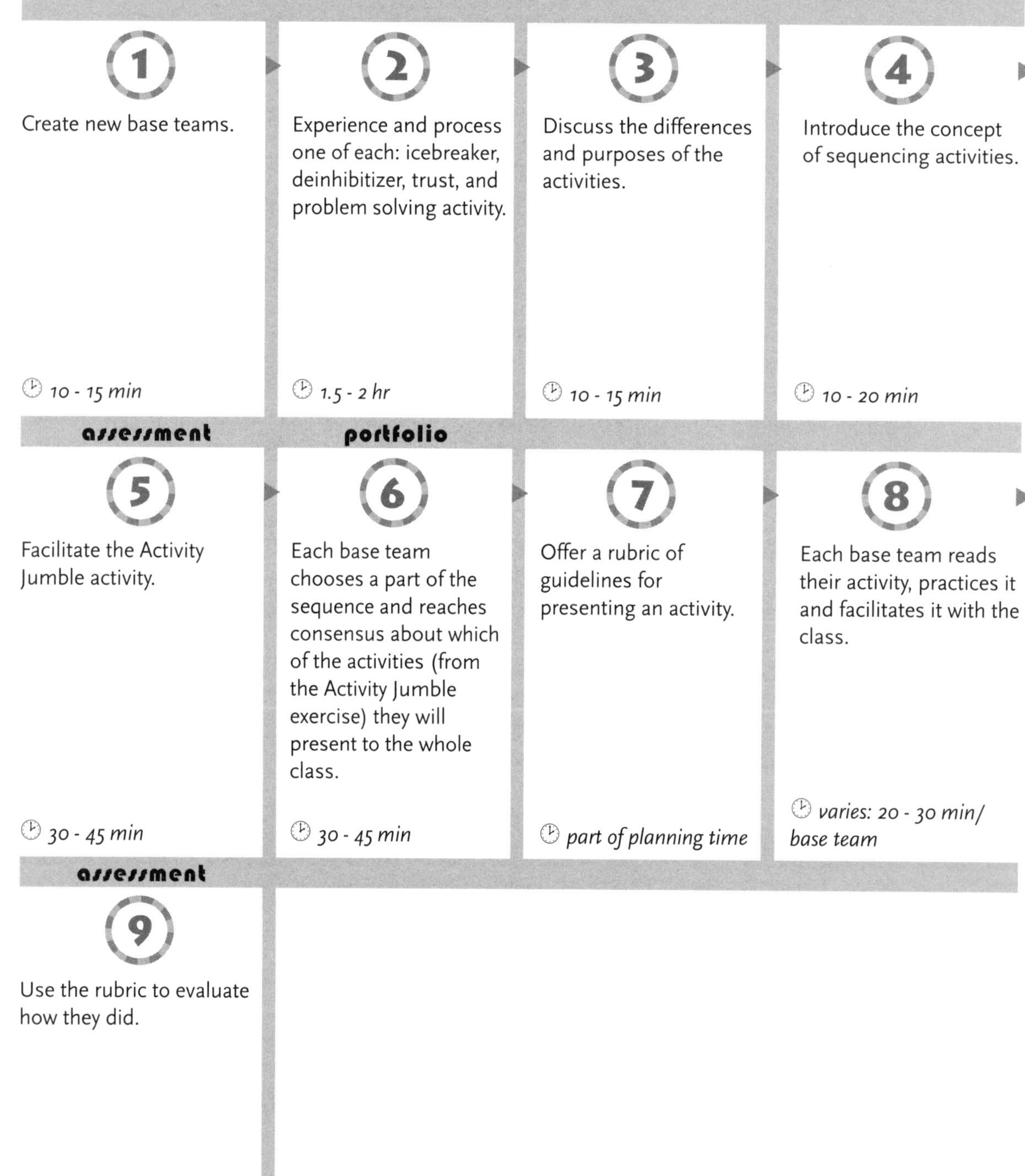

Lesson 2.A Sequence: Creating Community

2.A sequence

Create new base teams.

Base teams are groups that stay together over time, providing continuity. To create the teams, ask students to secretly write down three to five other people in the class with whom they would like to be on a base team. Collect the cards and create the teams based on who students put on their cards. Tell them that they will be on a base team with **at least one person** they wrote on their cards. Students will not always work in these teams, but will discuss deeper issues and do larger projects together. It is nice to mix things up a bit so that students are not always with the same people.

Experience and process one of each: icebreaker, deinhibitizer, trust, and problem-solving activity. The purpose of doing one of each of these activities with the class is to be transparent about sequencing activities when trying to build relationships. This is one step toward helping young collaborative leaders understand that 1) intentionally building relationships is important, 2) doing activities together is one way to intentionally build relationships (see Unit 3 to explore other ways), and 3) building relationships takes thought, and a sequence can be important when using activities.

See the Activity Sequence Chart from sequence 5 of 2.A for a generally accepted sequence for activities.[1] Please note that collaboration, leadership, and conflict resolution take front-and-center later in the sequence—during problem-solving. These are more complex concepts/issues that are better dealt with on a foundation of relationships that value cooperation, communication, and trust.

Do each of these activities with the class. After processing what occurred during the activity, note to the class what type of activity it is and why it is used. Have them quickly note the similarities and differences between the activities so that they can become more aware that all activities are not equal, and that there is a purposeful sequence.

activity

Circle Scramble

Focus: Learning about others, name game
Materials: None
Time: 10 - 15 minutes
Sequence: Icebreaker
Sources: See "Name by Name" in *Quicksilver* by Rohnke & Butler.

Suggested Procedure

1. The group stands in a circle.
2. Each person says his/her first name so that everyone can hear it.
3. Without any type of communication (verbal or nonverbal), group members must attempt to get in alphabetical order.
4. Once they think they have it, have them say their names to see how close they came.
5. Then go to round two (with no communications), attempting to fix the mistakes.
6. Go to further rounds, if necessary.
7. Try this with last names, birthdays, address numbers, etc. Anything that is sequential can work. The more people the better!

Sample Processing Questions

- How well did we do in this activity? How many rounds did it take?
- What strategies did you use to figure out where you fit in the circle?
- What was our common goal (vision) for this activity? Did we all know what it was?
- Did it help to know what the goal was, even though we were not allowed to communicate?

Facilitation Notes

This activity usually does not need much processing. It gives people a chance to work together and learn some things about each other in a non-threatening manner.

activity

Toss 10

Focus: Cooperation, communication, not being "perfect," perseverance
Materials: One soft throwable object for each person
Time: 20 - 30 minutes
Sequence: Deinhibitizer
Sources: Laurie Frank first learned this activity from Sarah Shatz. See *Journey Toward the Caring Classroom* by Frank.

Suggested Procedure

1. Have the group stand in a circle and give each participant a throwable object.
2. Tell the group that as you count to 10, they will do something with the object for each number.
3. Practice each number with the motion that goes with it. Here is a possible example:
   1 - Throw object in air and catch it
   2 - Throw object in air, clap once, and catch it
   3 - Throw object in air, clap twice, and catch it
   4 - Throw object in air, touch toe or knee, and catch it
   5 - Throw object in air, spin around, and catch it
   6 - Pass object to person on the right
   7 - Pass object to person on the left
   8 - Pass object to person across the circle
   9 - Throw object in air, do a little dance, and catch it
   10 - Throw object in air, throw arms over head, yell "Hey ho, way to go," and catch it
4. After the pattern is set and practiced, introduce a second part to the activity: counting to 10 as a group.
5. No planning is allowed, and someone in the group must call each number in order. When the number is called, everyone does the motion that was established for that number.
6. Here's the catch: if two or more persons says a number at the same time, then all must start over.

Sample Processing Questions

- Did you feel self-conscious when you dropped your object? Why or why not?
- How did we solve this problem? Did we communicate in any way? How?
- How was your frustration level? What made you want to keep try (or not)?
- What strategies did you personally use to help achieve the task?

Facilitation Notes

This activity is really two activities combined into one: first the pattern of throwing and catching the object, and then the count to 10 activity. Interestingly enough, many groups get through the counting to 10 on the first try. If this happens, ask them to repeat it. Many times the second time is more difficult.

Assess the frustration level of the group. If they are unable to get to 10 and have to keep starting over, they may become overly frustrated. This activity can be a good assessment of a group's problem-solving abilities and can help you decide if they are ready to move on to other, more challenging tasks, or continue with more ice-breaker/deinhibitizer activities.

activity

Willow in the Wind

Focus: Risk taking, physical/emotional trust, trustworthiness, empathy
Materials: None
Time: 30 - 45 minutes
Sequence: Trust
Sources: Originally found in More New Games by Fluegelman. See also *Adventure Education for the Classroom Community* by Frank & Panico, *Changing the Message* by Albin, "Circle of Friends" in *Diversity in Action* by Chappelle & Bigman, "Circle Pass" in *Games for Group I* by Cavert, and *Journey Toward the Caring Classroom* by Frank.

Suggested Procedure

1. You will need space for one, two or three groups—the hallway, school lawn, gym or all-purpose room works best. A classroom is possible, but tight.
2. Get into groups of 8 to 10 and ask for one group to volunteer as models. Show the whole activity before having everybody do it.
3. One person in the small group is the "faller." This person assumes **falling position**:
   - Stand with feet together
   - Cross arms over the chest
   - Hold body as stiff as a board (tell him/her to squeeze his/her bun cheeks—not only is this good exercise, but it keeps his/her body from bending)
4. The other people are spotters and stand in a tight (shoulder to shoulder) circle around the faller. They assume **spotting position**:
   - Stand shoulder-to-shoulder
   - One foot in front of the other to provide a stable base (it doesn't matter which one)
   - Knees bent
   - Hands up, palms out, with fingers pointing toward the sky
5. Explain the **communication** between faller and spotters:
   - Faller: "Spotters ready?"
   - Spotters: "Ready!"
   - Faller: "Falling!"
   - Spotters: "Fall on!"
6. At this point, the faller tips backward into the waiting hands of the spotters, who gently push the faller to another part of the circle. This continues until the faller chooses to stand and says, "I'm on my own."

7. There should be at least three hands on the faller at all times so that nobody has to hold the entire weight of the faller by his/her self.
8. Remind the spotters not to turn their hands so that their fingers face the floor as this is poor leverage and does not work well.
9. If something should go wrong (faller bends or tilts to the side) the spotters' responsibility is to keep the faller's head, neck and shoulders from touching the floor.
10. Only volunteers go in the middle, and individuals can only volunteer themselves!

Sample Processing Questions

- How did it feel to be surrounded by so many people during this activity?
- Did you feel like the spotters were taking care of your safety? How could you tell?
- Compared to other leaning activities, how was this one?

Facilitation Notes

Remind students to monitor how hard they are pushing people. **Discontinue this activity if participants roughly push those in the middle, and do not change the practice with reminders.** Remember people can break trust as well as build it. The activity, in itself, does not engender trust between people—what we **do** with the activity determines whether trust is built or broken.

Willow in the Wind can be difficult for some because of all the touching involved. Make sure people know they have a choice about whether or not they go in the middle. It can help to acknowledge that touching is an issue for some people. It may be useful to have an adult join each of the groups for the first couple of tries until students understand what is expected of them.

Encouraging people to close their eyes offers a different dimension to the activity.

activity

Unification

Focus: Thinking outside the box, sharing space, leadership, working together
Materials: 3 to 12 feet lengths of rope
Time: 20 - 30 minutes
Sequence: Problem Solving
Sources: See "Jumping Stars" in *Adventures in Peacemaking* by Kreidler & Furlong, "Lily Pads and Islands" in *Affordable Portables* by Cavert, "Cliques" in *Games for Change* by Dodds & Prosser-Dodds, "Fusion" in *Journey Toward the Caring Classroom* by Frank, and "Mergers" in *Quicksilver* by Rohnke & Butler.

Suggested Procedure

1. Give each member of the group a length of rope and asked them to tie the rope into a circle using any type of knot.
2. Then have participants put their circles on the ground and stand so that their **feet are entirely within the circle**.
3. On a signal, everyone in the group must move to another circle. When done, everyone must have his/her feet entirely within the new circle.
4. The catch is that each time the group moves, one circle is removed. Still, they must all fit their feet within an available circle—no one is ever eliminated.
5. Keep a larger circle in play for the last attempt (see facilitation notes for clarification).

Sample Processing Questions

- When did you realize that you would need to share circles? Did you feel it might be necessary to eliminate people instead?
- If this was an elimination exercise, how might it have been different?
- What do you think are the values of competition?
- When might competition get in the way of accomplishing tasks?
- In your life, where do you think competition is appropriate and where might it be detrimental?
- When did you know that everyone would fit inside the circles?
- When the group all fit, how did you feel?
- How did you work together to make this successful?

Facilitation Notes

There are two main realizations a group must come to in this activity. One is that they must share circles. The other is that their whole body does not have to fit in the circle, only their feet. This means people can sit on the ground outside of the circle, placing only their feet in the circle. At the end, everyone will be crowded around one circle.

Make sure a larger rope is available to finish the activity. It is not, however, necessary to leave the biggest circle as the final one. The more challenging the task, the more the group must strategize.

This is a good activity to end a group's time together, or if participants are competing with each other within the group. It serves as a useful metaphor for joining together and sharing space.

Discuss the differences and purposes of the activities.

What made each of the activities unique compared to the others? What do you think was the purpose/focus of each activity? What do you think each is designed to do?

Why even bother to do activities such as these? What do they accomplish? What is the point? Why the sequence? What if we had done them in a different order? Would it have made any difference? Imagine we were a group coming together for the first time? How would it feel to do the Willow in the Wind activity first?

Introduce the concept of sequencing activities.

There is general agreement in the field of adventure education that activities must be sequenced to help facilitate the development of group cohesion. Christian Bisson recently explored a "hypothetically correct sequence" (HCS) for his doctoral dissertation, concluding that the "HCS appears to be effective at developing group cohesion among participants... [and] the HCS seems to be a viable progression to follow" (1997, unpublished dissertation).

The hypothetically correct sequence focuses on **group formation**, moves to **group challenge**, and then finishes with activities that require **group support**. It is important to note that the HCS follows the general rule that activities start out as low threatening icebreaker/acquaintance activities and progress through activities that require more commitment and trust from group members. Finally, group members are asked to support and be supported by others as they attempt problem solving challenges together.

Following is a recognized sequence with descriptors at each level:

Activity Sequence Chart

Ice Breakers

- Introduces participants to each other
- Teaches names through games
- Relaxes people through fun
- Builds comfort by including people

Deinhibitizers

- Expects more of people
- Breaks barriers
- Helps people begin to make choices about risk taking
- Stresses acting silly, being put on the spot, touching

Trust Activities

- Explores risk taking
- Looks at idea of being trustworthy
- Helps people decide when, how and where to trust others
- Gives people the opportunity to trust and be trusted

Problem-Solving Initiatives

- Builds on foundation of emotional trust
- Explores issues of leadership and followership
- Gives people the opportunity to learn about solving problems collaboratively
- Provides the opportunity to resolve conflicts within the group peacefully

Facilitate the Activity Jumble activity.

activity

Activity Jumble

Focus: Sequencing activities
Materials: Activity Sequence Chart from the previous step (sequence 4 of 2.A, above), Activity List and Descriptions (p. 160-161)
Time: 30 - 45 minutes
Sequence: Problem Solving

Suggested Procedure

1. Divide the class into groups of 3 to 5.
2. Give each person an Activity Sequence Chart and an Activity List and Descriptions.
3. Have students work together in small groups to categorize the activities using the chart and the descriptions. Explain that there are no wrong answers as long as they can provide a rationale for their decisions.
4. When done, ask one group to share the activities they chose as icebreakers. Have groups compare their lists. Ask them to provide a rationale for why they chose a different activity.

5. Do the same for deinhibitizers, trust, and problem-solving.

Sample Processing Questions

- What did you learn about sequencing activities?
- Why do you think it's important to know the difference between activities?
- When might you choose to do an activity with a group?

Facilitation Notes

This activity begins the process of teaching students to facilitate. Learning how to choose and facilitate activities is a useful skill when leading from within a group, as it gives young people some tools to use when needed. Knowing activities is not very useful unless one knows when to use them.

In adulthood, it may not be practical to do a physical activity with a group. There are other ways to break the ice, build trust and create teams, however. That is why it is important to understand the concepts that underlie the reason for choosing one activity over another.

Although there are no real right or wrong answers, here is Laurie Frank's opinion (and rationale) on how she would categorize activities:

- **A What?:** Deinhibitizer (puts people on the spot, can also be seen as a problem-solving activity if trying to figure out how to get the objects all the way around)
- **All Aboard:** Problem-Solving (requires trust, but also involves figuring out a problem)
- **Are You More Like?:** Icebreaker (low threatening way to learn about others as people have a choice about divulging information)
- **Channels/Pipeline:** Problem-Solving (a direct problem to solve; leaders usually emerge during this one)
- **Commonalties:** Icebreaker (gives people a low-threatening way to learn about each other)
- **Don't Touch Me:** Problem-Solving (another direct problem where people must listen to ideas and try to put the ideas into action)
- **Human Bingo:** Icebreaker (low threat activity about giving people an opportunity to learn about each other)
- **Name Tag:** Icebreaker (another low threat activity about hearing names and giving people a chance to get their voices out there, can also be a very low level problem-solving activity if the focus is on trying to beat the time)
- **People to People:** Deinhibitizer (requires touching)
- **Screaming Toes:** Deinhibitizer (acting silly in the group, puts people on the spot)
- **Search and Rescue:** Trust (whenever anyone is asked to close his/her eyes or put on a blindfold, consider it a trust activity; challenge students' rationale if they categorized this as an icebreaker or deinhibitizer)
- **Warp Speed:** Problem-Solving (although fairly low threatening, it does require leadership, listening to ideas, and putting the ideas into action)
- **Yurt Circle:** Trust (requires people to put their physical and emotional well-being in the hands of others)

Activity List and Descriptions

As a group categorize each activity as either an icebreaker, deinhibitizer, trust activity, or problem solving initiative. Use the Sequence Description Chart as a guide.

A What?

Sit in a circle. Start an object around by turning to the person on your right, handing them the object, and saying, "This is an aardvark." They respond by saying, "A What?" Then you say, "An Aardvark." They turn to the third person and say, "This is an Aardvark." When the third person says, "A What?" the second person turns back to you and says "A What?" This pattern continues. Next start a different object going to the left: "This is a gorilla." "A What?" "A Gorilla." And so on....

All Aboard

The entire group is asked to stand on a tarp or platform. After they are able to do that, the tarp is folded (or a smaller platform introduced) until it is difficult to accomplish the task.

Are you More Like?

Designate two sides of the room. Ask people to put themselves on one side or the other according to questions you ask. Have them discuss why they put themselves there either as a group or with those around them. Example: Are you more like chocolate or strawberries?

Channels/Pipeline

Everyone is given a channel (1/2 inch PVC pipe cut lengthwise). They must move a marble without touching it or walking when they have it. In other words, they must pass it by using their channels. If the marble hits the ground or touches someone, they must start over. A nice sequence is to have people work in groups of three to practice, then form a group of 10 or so and pass it in a circle. Finally, they must move the marble across the room as a large group.

Commonalities

The larger group divides into smaller groups of two people or more. They are asked to generate a list of things that are common to all the people in the small group that cannot be seen (i.e., they were all born in the same season, not everyone wears glasses).

Don't Touch Me

This is a timed event. Everyone identifies someone across the circle from him/her. The object is to trade places with that person without touching anyone in the process. A circle is placed in the middle, and everyone must touch in that circle once during the move (this counteracts the "clock" phenomenon). See if they can do better on their time.

Human Bingo

Give each student a bingo sheet. The object is to get as many different people to sign their card as possible. They are looking for people who are contrasting matches to them for each category.

Name Tag

This is a timed activity where everyone says their name, in order, as fast as they can. Then the group attempts to beat their time. Try it the other way, reversing the order around the group, too.

People to People

Have everyone get a partner (see "Puzzle Pairs" as a way to do this). You do not have a partner, and you are the caller. Call out "people to people," and have them repeat it back. Then call out two body parts (e.g., hand to hand). The partners touch these body parts together. Call out two more body parts. When you call out "people to people" a 2nd time, they find a new partner. (For younger children, they can stay with the same partner. For older children, you can find a partner, which will then give them a new caller.) For challenge by choice, teach the extension: thumb and pinky out, while curling in the rest of the fingers. That way they can choose to use it if they don't want to get too close.

Screaming Toes

Stand in a circle. Everyone looks up, then looks down, then at each other. One must look at one person and keep looking, no changing in midstream. If the person being look at is looking at someone else, then the looker is safe. If eye contact is made, then the two are zapped. They each scream, and the group moves on to the next round.

Search and Rescue

One person is blindfolded and has a sighted partner as a guide. The guide throws an object and verbally directs the blindfolded person to the object. The blindfolded person then picks up the object and takes it to the "safety" zone. Once deposited, they switch roles.

Warp Speed

Start like group juggle, but use only one object. Once the pattern is established, time how long it takes to get through it once. Then ask if they think it can be lowered and how. State that the only rule is that it must touch everyone in the same order. Challenge the group to do this in under 10 or 5 seconds. Once they realize they can move themselves, things pick up.

Yurt Circle

You need an even number of people. Stand in a circle. Count off by two's so that every other person is the same number. Have the ones turn around. Everyone holds hands. On the count of three everyone gently leans forward to hold the circle up with tension. Try it with everyone leaning back. For a bigger challenge have everyone face to the center of the circle. Have the ones lean forward while the twos lean back. Switch.

Each base team chooses a part of the sequence and reaches consensus about which of the activities (from the choices that follow) they will present to the whole class.

Have groups choose a category or assign them randomly so that at least one group presents an icebreaker, a deinhibitizer, a trust, or a problem-solving activity. The small groups then choose one of the activities in their assigned category to facilitate with the whole class. Only the activities that have not been presented previously are included below. Refer students to the facilitation on p. 135.

activity

ARE YOU MORE LIKE?

Focus: Getting to know each other, sharing information
Materials: None
Time: 10 - 20 minutes
Sequence: Icebreaker
Sources: Laurie Frank first learned this from Chris Cavert. See *Games for Teachers* by Cavert & Frank.

Suggested Procedure

1. Clear the desks or tables away and ask people to stand in the middle of the room.
2. Explain that you are going to give them two choices that can be interpret any way they want.
3. Ask the group, "Are you more like chocolate (point to one side of the room) or strawberries" (point to the other side of the room).
4. Have them move to the side of the room that is most like them. Again, they can interpret it any way they want. Some may simply like strawberries more than chocolate. Others my choose strawberries because they are healthier than chocolate and they see themselves as healthy people.
5. Ask them to take a minute and tell those around them why they chose that side.
6. Share one or two reasons with the whole group.
7. Repeat the process. You can either create your own questions or use the following: Are you more like: • chocolate or strawberries? • a carpet or a wood floor? • a hard cover or paperback book? • skim milk or 2%? • a chair or a couch? • apples or oranges? • pants or shorts? • a bath or a shower? • jeans or khakis? • glass or plastic? • a bunk bed or a twin bed? • a bus or a plane? • a bracelet or a necklace? • solids or stripes? • sandals or shoes? • shade or sun? • cursive or printing? • buttered or plain popcorn? • potato chips or pretzels? • a run or a walk? • soda or water? • panic or relaxed? • sky diving or sun bathing? • a day at the beach or a day white water rafting? • fact or fiction? • a cat or a dog? • a chocolate or an oatmeal cookie? • the escalator or the stairs? • fish or steak? • cheese or pepperoni? • an amusement park or a water park?

Sample Processing Questions

- Did you learn anything about yourself compared to other people?
- Were you surprised by anything you heard?
- What does this activity tell us about ourselves as a class?

Facilitation Notes

As stated above, make sure people know they can interpret the questions any way they want. People can then see there are many ways to interpret things, and they can still have their own point of view. Stress that there are no wrong answers.

SAFETY ISSUES: Remind everyone about the right to pass. They don't have to answer any particular question and can choose not to answer any or all of them.

activity

COMMONALITIES

Focus: Getting to know each other
Materials: Paper and markers
Time: 15 - 30 minutes
Sequence: Icebreaker
Sources: *Quicksilver* by Rohnke & Butler

Suggested Procedure

1. Have the large group get in smaller groups of 3 to 5.
2. Give each small group a large piece of paper and some markers.
3. Begin by asking the large group, "What things can you see that we have in common?"
4. Have people suggest a few.
5. Then ask them to find out what each of them has in common in their small groups that cannot be seen. It is necessary for them to ask each other questions to learn what these are.
6. After 5 to 10 minutes, ask each group to share their top 3 to 5 favorite commonalities with the large group.

Sample Processing Questions

- What did you learn about other people in your group?
- Would you have learned these things in any other way?
- Why is it important to learn about people you work with?

Facilitation Notes

This activity can help to bring out the issues of stereotyping and begin breaking down barriers between different peer groups.

SAFETY ISSUES: Remind people that they have the right to pass and to share only what they are ready to share.

activity

PEOPLE TO PEOPLE

Focus: Sense of humor, being put on the spot, touching, mixing
Materials: None
Time: 10 - 20 minutes
Sequence: Deinhibitizer
Sources: Laurie Frank first learned this from Pete Albert. See "Student to Student" in *Adventure Education for the Classroom Community* by Frank & Panico, and *Journey Toward the Caring Classroom* by Frank.

Suggested Procedure

1. Have everyone get a partner (you do not have a partner, you are the caller).
2. Call out "people to people," and have them repeat it back.
3. Call out two body parts (e.g., hand to hand) and have partners touch these body parts together.
4. Call out two more body parts and have partners touch those body parts together.
5. When you call out "people to people" a second time, they find a new partner.
6. Try this a couple of times. Then add a new rule: The next time, you (the caller) will also find a partner. That way someone will be left without a partner and will become the new caller.
7. For challenge by choice, teach the extension: thumb and pinky out, while curling in the rest of the fingers. The thumb touches their own body part and the pinky is held out to touch the pinky of their partner. Students can choose to use it if they don't want to get too close.

Sample Processing Questions

- How did you like getting a different partner each time? Was it easy or hard?
- Did you have fun playing this game? Why or why not?
- What can we do to make it even more fun?
- Did you choose to use the extension? Was that okay?

SAFETY ISSUES:

- Remind students of challenge by choice and the right to pass. Some people do not like to be touched, and they should have a graceful way to opt out if they need to.
- Introducing the concept of the extension (#7 above) gives people a graceful way to handle it, and it is good information for processing after the activity.

activity

Screaming Toes

Focus: Breaking down barriers by acting silly in the group
Materials: None
Time: 10 - 20 minutes
Sequence: Deinhibitizer
Sources: Laurie Frank learned this activity from Steve Butler. See "Making Connections" in *Adventure Education for the Classroom Community* by Frank & Panico, and *Journey Toward the Caring Classroom* by Frank.

Suggested Procedure

1. The group stands in a circle.
2. On a signal given by you (the caller) everyone is asked to look down at someone's toes (not their own), then (on another signal by you) to look up at that person.
3. If they do not make eye contact (i.e., the person they are looking at is looking at someone else), they do nothing.
4. If they make eye contact (i.e., they are looking at each other), they let out a short scream or yell. Try this for 10 to 15 rounds.
5. If the group is over 10 to 12 members and few people are making eye contact, here is a variation: Divide into two smaller groups. Every time a person makes eye contact with someone they scream and then move to the other circle.

Sample Processing Questions

- Was this activity risky for you? Why or why not?
- Why might it be important to be able to laugh together as a group?

Facilitation Notes

Sometimes it is difficult for people to relax in a group setting, thus limiting their risk-taking abilities. This activity is a hands-down favorite and encourages group members to laugh with each other. Later, when engaged in more intense tasks, the ability to laugh with the group, and at oneself, can be invaluable.

This activity can be used as a tension-reliever and energizer at any time. The only time this activity may not work is during the first hours of a group's existence, when people are still trying to get to know each other, and before the ice has been broken.

SAFETY ISSUES: Remind people to not scream loudly in someone's ear or face.

activity

Search and Rescue

Focus: Physical and emotional trust, trustworthiness, empathy, risk taking
Materials: Eye coverings (optional), soft throwables (like wadded up paper or stuffed animals)
Time: 15 - 20 minutes
Sequence: Trust
Sources: See "Evidence Rescue" in *Adventure Education for the Classroom Community* by Frank & Panico, *Journey Toward the Caring Classroom* by Frank, and "Rabid Nugget Rescue" in *Silver Bullets* by Rohnke.

Suggested Procedure

1. Create a "safety zone" by marking an area or by placing a box/bag/trash can in the middle of the room.
2. Have everyone get a partner, and give each pair a soft throwable item.
3. The items are described as "people who have been lost in the wilderness." The object is to throw the item, retrieve it, and get it into the "safety zone." It must be done in this way:
   - One person in the pair covers or closes her eyes.
   - The partner (the "guide") throws the object a distance from the safety zone. The guide then uses only verbal instruction to lead his partner to the object. The guide may touch his partner only if an obstacle is encounter and safety becomes a factor.
   - Once the partner gets to the item, the guide has the partner pick up the item, and then the guide verbally leads the partner back to the safety zone.
   - Once the item is back to safety, the two switch roles and repeat the process.

Sample Processing Questions

- Was it easier for you to be the guide or the retriever? Why?
- What did you (could you) do as the guide to help your partner feel comfortable and establish trustworthiness?
- What were the communication issues that you experienced? How did you handle them?
- Did you feel safe during this activity? Why or why not?

Facilitation Notes

Since the person who cannot see is relying on the partner who can see for safe passage, some notable issues arise around trustworthiness, communication, and perspective taking. Many times an object is thrown under a desk, and the guide leaves his partner standing there alone without telling her that he is leaving or why he left. This is a communication issue that has ramifications around building trust and establishing trustworthiness. It is also an issue about perspective taking and how it might feel to be standing there, vulnerable, not knowing what is going on.

SAFETY ISSUES:

- When an object is thrown under a table, bench, or desk, allow the guide to move it away from the barrier so that the retriever does not have to crawl around on the floor.
- Demonstrate for the group what can happen if an object is thrown next to a wall and the guide tells his partner to bend over to pick up the object without warning her about the wall.
- Remind participants to move slowly. Speed can be dangerous, especially in a room filled with furniture. Although the person moving fast may not mind the speed, it puts others at risk.

activity

ALL ABOARD

Focus: Decision making, leadership
Materials: A tarp
Time: 20 - 30 minutes
Sequence: Problem Solving
Sources: See *Adventure Education for the Classroom Community* by Frank & Panico, "Box Top" in *Affordable Portables* by Cavert, *Journey Toward the Caring Classroom* by Frank, *Silver Bullets* by Rohnke, and *Teamwork and Teamplay* by Cain & Jolliff.

Suggested Procedure

1. Clear the desks or tables away and stand in a circle around an open tarp.
2. Tell the group that they are scientists who are stuck in a huge pit. Although still in its testing phase, their only hope is to get everyone onto the growth machine pad together. Once they have enough weight (the weight of the entire team) on the pad for five seconds, they grow one size bigger. This must be done over and over again, until they are big enough to get out of the pit.
3. Start with the tarp open wide. When they get the entire group on the tarp, count to five. Then ask them to step off.
4. Fold the tarp by a third. Try this again.
5. Continue folding and standing until it is a bit of a struggle. Then fold it one more time—just a bit.

Sample Processing Questions

- How did your strategy change during this activity? Why did it change?
- What did you do to listen to each other so that you could make a group decision? Did it work?
- Were you a listener, a talker, or both during this activity? How do these enter into being a leader?

Facilitation Notes

As the tarp gets smaller and smaller, the solution to the problem changes. At first, students can simply step on the tarp. As the space tightens, it is necessary to coordinate their efforts.

SAFETY ISSUES:

- Make sure that you are spotting this activity. When the tarp gets smaller and people need to hold onto each other to stay on it, stand around the edge and be ready to stop someone from falling.
- Warn students to step off the tarp slowly, rather than in a large, uncontrolled clump.
- If the tarp ends up so small that they are unable to get everyone on it after a few attempts, stop the activity, tell them you are giving them a "shrinking pill," and open the tarp a little.

activity

DON'T TOUCH ME

Focus: Group goals, decision-making, leadership
Materials: A hula-hoop or short rope, stopwatch
Time: 15 - 30 minutes
Sequence: Problem Solving
Sources: Laurie Frank first learned this activity from Cindy Simpson. See *Adventures in Peacemaking* by Kreidler and Furlong, *Diversity in Action* by Chappelle & Bigman, "Centerpiece" in *Games for Teachers* by Cavert & Frank, *Journey Toward the Caring Classroom* by Frank, and *Quicksilver* by Rohnke & Butler.

Suggested Procedure

1. Clear the desks or tables away and stand in a circle. Place the hula-hoop on the floor in the middle of the circle.
2. Ask everyone to identify a partner across the circle from him or her. Have them point to the feet of their partner—they should be pointing at each other's feet.
3. The object of this initiative for the partners to trade places without touching anyone else. At some point in the switch, everyone must put his/her foot in the hula-hoop. This can be done together or alone, it is up to them.
4. This is a timed activity. The time will start when the first person moves from his or her spot, and the time will stop when the last person has assumed his or her place.
5. Try this a number of times, allowing for some strategizing between attempts.

Sample Processing Questions

- Did you set a group goal for this activity? What was it?
- How did you arrive at this goal?
- Can you think of other solutions for this task?

Facilitation Notes

As with many of these initiatives, there are a variety of ways to accomplish the task. Some groups remain in the circle formation, while others rearrange themselves to improve their efficiency. Most groups will choose to set a goal to lower their time, but a few have chosen to see how many different solutions they can try.

SAFETY ISSUES: Remind students that they are going to be moving near other people and to make sure not to move so fast that they bump into others.

Offer a rubric of guidelines for presenting an activity.
Have students study this in their group before presenting their activity.

Criteria	-	o	+	Bonus
Gets the groups attention before starting	Gives directions without getting people's attention	Gets people's attention, but gives directions before everyone is ready to listen	Gets people's attention and waits until everyone is ready	Has a strategy besides yelling to get people's attention: "If you can hear me clap once," or signals a "time out."
Gets the group in a place where everyone can see and hear (circle, line, etc.)	Does nothing to situate group so that everyone can see and hear	Gathers people but allows them to sit or stand in a clump	Asks for, and waits for, people to get into a circle or a line so that everyone can see and hear	When people stand in front of others, reminds them to make space so that everyone is included in the circle.
Activity Introduction	Gives long and rambling directions that are not sequential and leaves out rules or directions.	Gives sequential directions for the activity that include all the rules.	After giving directions, asks if there are any questions.	Gives directions **and** shows them.
Processes Activity	Does not process the activity at all.	Asks questions for participants to answer about the activity.	Asks follow-up questions when an issue is brought up by a participant.	In addition to talking, uses a way to process that is visual, physical, or artistic.

Each base team reads their activity, practices it, and facilitates it with the class.
As students present their activities be alert for safety issues. If they do not cover safety in their directions, remind them to do so. If you see a safety issue occurring, be prepared to step in and stop the activity until the safety issue can be corrected.

Use the rubric to evaluate how they did.
Have each small group fill out a rubric form for themselves. You can also have yourself and some of their peers fill out a rubric to give them more feedback.

Lesson 2.B Overview: Compose a Personal Definition of Leadership

Focus: Practice reflection

1	2 (portfolio)	3 (assessment)	4
Facilitate the Think-Pair-Share activity about someone you admire as a leader.	Write a paper about why this person is a leadership role model for you.	Trade papers with three other people and discuss.	Group Brainstorm: What leadership qualities does this person have that you admire?
15 - 20 min	1 - 1.5 hr	45 min - 1 hr	15 - 20 min

5	6	7 (assessment)	8 (assessment)
Read the summary of life skills that support leadership.	Journal assignment.	Have students choose one life skill and write a SMART goal in their journal to work on that life skill during the next month.	Journal assignment.
1 - 1.5 hr	10 - 15 min	15 - 30 min	20 - 30 min

Time Frame: 4 - 6 hours

Lesson 2.B Sequence: Compose a Personal Definition of Leadership

Facilitate the Think-Pair-Share activity about someone you admire as a leader.

activity

Think-Pair-Share

Focus: Listening, perspective taking
Materials: None
Time: 15 - 20 minutes
Sequence: Icebreaker
Sources: Laurie Frank first learned about Think-Pair-Share from Mary Henton.

Suggested Procedure

1. Have students get into pairs.
2. Ask each participant to think about someone s/he admires as a leader. It can be a personally acquaintance or someone s/he has never met—someone famous, a family member, friend, someone alive or dead. Give the group a moment to get someone in their minds.
3. Explain that they will have one minute to silently think about this person and what it is about their leadership that they admire. Time them for one full minute.
4. Now tell them that they will have 2 minutes to tell their partner about the person they are thinking of and what they admire about that person. One person will talk for their two minutes, and the other person may only listen and not say anything.
5. Have them choose who will go first in their pair, then time them for 2 minutes as one partner describes the person s/he admires.
6. After two minutes, call time and ask them to switch roles, giving the other person 2 minutes to talk.
7. After the 2 minutes are up, give them 3 minutes to discuss anything else about their person (both people can talk now).
8. If they continue to engage in conversation, give them more than 3 minutes.

Sample Processing Questions

- Was it difficult to choose someone you admired or did one person jump right into your mind?
- What made you choose the person you did?
- What traits are most important to you in leadership?
- Do you see any of these traits in yourself?
- How did it feel to have someone's undivided attention when you were talking?
- How was it to be unable to respond to someone when they were talking?

Facilitation Notes

It can be difficult for people to sit silently, even for a minute. The group will usually settle down after the initial bout of discomfort. If there is talking or giggling, simply extend the time a little. Talking to someone who is listening intently can be uncomfortable for some. People may not give full attention because they are thinking about what they want to say. Encourage students to be fully present when listening. Encourage those who are talking to speak spontaneously rather than trying to craft what they might want to say. The spontaneity can bring up some surprising insights for the speaker and talker alike. Earlier activities and experiences together will probably help students feel more comfortable than if you do this activity earlier in the sequence.

Write a paper about why this person is a leadership role model for you.

Go back to the person they chose as a leader role model. Ask them to write a 1 or 2 page paper about why they chose that person.

Trade papers with three other people.

Have students make three copies of their papers. In small groups of 4, have them trade papers. After they read them all, hold a discussion about the papers: What leadership traits were similar, different? Are any of these traits more important than others? If so, why? If not, why not?, etc.

Group Brainstorm: What leadership traits does this person have that you admire?

Ask students to think about the leadership traits talked about and to call them out as you write them on an overhead or flip chart paper. Ask if they see any patterns or notice anything interesting.

Make the statement: "Nobody has all of these traits, but everybody does." Discuss what that means to them. If it hasn't been said, note that individuals have many strengths, but cannot have them all. If we pool our strengths through collaboration, there's a good chance that we, in this room, can cover most or all of these traits.

Read the summary of life skills that support leadership.

There are many life skills or character traits that support leadership. Life skill development is important for young people because it helps them focus on who they are, and how they interact with the world around them. Although 13 life skills are discussed here, there are many more that could be added. Following you will find a paragraph about each of these life skills:

- Caring/Compassion
- Courage
- Curiosity
- Excellence
- Flexibility
- Growth
- Integrity
- Passion
- Patience
- Perseverance
- Resourcefulness
- Sense of Humor
- Service

Using the jigsaw approach, break the class into 3 to 5 small groups and assign each group 2 to 4 of the life skills that support leadership. Have them read the corresponding paragraph, discuss them, look up their definition, do some further research in the library or on the Internet, and then figure out how to describe them to other people. In essence, they become the experts on their assigned life skills.[2]

When ready, create new small groups so that there is at least one person from each of the "expert" groups in the new groups. The experts then teach the others about the traits they studied.

Summary of Life Skills That Support Leadership

There are many life skills and character traits that support leadership. Here are 13 that can get you started:

Caring/Compassion: Showing concern for others; wanting to help others.

Caring and compassion grow from empathy—being able to understand what another person is feeling even if you don't feel it yourself. More than sympathy, which is to feel sorry for someone, caring and compassion are a gut-level feeling that causes action. When someone shows true caring and true compassion for someone or something, he or she does something about it. It may be as simple as listening to a friend who is worried, or as involved as helping to end world hunger.

Our task must be to free ourselves by widening our circle of compassion to embrace all living creatures and the whole of nature and its beauty. —Albert Einstein (1879-1955, physicist)

Courage: The ability to face difficult situations.

Without fear, there can be no courage. Everyone has to deal with danger, conflict, uncertainty, and pain. Life is full of risk, both physical and emotional. Many of these situations may cause a person to want to run and hide. It is just at that moment when courage is needed to face the situation. In a leadership scenario, courage is needed when one's integrity is being challenged, (i.e., when friends are planning something that you know is harmful). At this point, it may be necessary to do the right thing rather than go along with the crowd.

With courage you will dare to take risks, have the strength to be compassionate, and the wisdom to be humble. Courage is the foundation of integrity.
—Keshavan Nair, Ph.D. (d. 2002, author, corporate executive, expert on leadership)

Curiosity: Wanting to know, get information, or learn about one's world.

The world can be an exciting place if you are interested in it. Curiosity is about asking questions and about wonder. It opens the mind. Learning is built on it, which means it is the foundation of growth. The more we learn the more we grow. Developing one's curiosity means that you are not satisfied with what you see and are interested in what you cannot see. The next time you look out the window, ask yourself questions about what you see outside. What kind of bird is that? Why is there only one? Does it stay in the winter or go somewhere else? When talking with a friend, ask questions and allow them to talk more than you. It's amazing what you may learn!

Don't let anyone rob you of your imagination, your creativity, or your curiosity. It's your place in the world; it's your life. Go on and do all you can with it, and make it the life you want to live.
—Mae Jemison (b. 1956, astronaut)

Excellence: The quality or art of being outstanding.

Excellence is not about being perfect—it's about striving to be the best one can be. It means following through on commitments, doing something to the best of your ability, and not being satisfied with second-rate work. Excellence is the difference between getting something done and doing it well. Excellence involves a host of other life skills, such as: effort, perseverance, integrity, and organization.

Excellence is to do a common thing in an uncommon way. —Booker T. Washington (1856-1915, educator)

Flexibility: The ability to change or alter plans when necessary.

Problem solving requires flexibility. Without it we easily fall into a rut of doing the same thing over and over, getting the same less-than-helpful result. Flexibility means that you may need to change your mind, or try something even if you think it may not work. It is important to note that being flexible does not mean being impulsive and acting without thinking. It does mean being open minded, letting go, allowing others to try things, and accepting that there is more than one way to do something. Being flexible means changing your focus from "me" to "we."

> *Flexibility and adaptability do not happen just by reacting fast to new information. They arise from mental and emotional balance, the lack of attachment to specific outcomes, and putting care for self and others as a prime operating principle. Staying open—emotionally—insures internal flexibility.*
> —Doc Childre & Bruce Cryer, From Chaos to Coherence (Institute of HeartMath)

Growth/Personal Growth: A process that produces change and progress.

Leadership is not a thing, but a never-ending journey. On this journey, it is necessary to pay attention to oneself and continue to grow in a variety of ways: with your head, your heart, and your spirit. Discover what you do well and do it better. Discover what you do not do well and develop it. Recognize that you are always learning and always changing. To do this well it is necessary to pay attention. Making mistakes is a great way to learn and to grow. Be ready to accept the mistakes in yourself and others—and grow from them.

> *I am always doing that which I cannot do, in order that I may learn how to do it.*
> —Pablo Picasso (1881-1973, painter and sculptor)

Integrity: To act according to what's right and wrong.

Integrity is about being true to yourself, your values, and what you believe in. It is a way to exercise personal power. Integrity is nothing unless it is acted upon. For example, if you know name-calling is wrong, then you do not call someone names. Above that, if you hear someone else calling names, to act with integrity means that you will speak to that person. Integrity means keeping promises, being honest, telling the truth, and following through. Sometimes it takes courage to act with integrity, and it is sometimes difficult to stay true to yourself. In the end, those who act with integrity earn the respect of oneself and others.

> *Our character [integrity] is what we do when we think no one is looking.* —H. Jackson Browne (author)

Passion: An enthusiastic interest in a subject or activity.

To be passionate about something means to believe in it deeply. It is easier to get something done if one is passionate about it. Others will soak up the enthusiasm and help with your cause. Without passion, it is easy to let things fall through the cracks, because it may not feel important. Passion is especially important when things are difficult. Amazing things can happen with the energy of passionate people. Think of people who were passionate about what they did: Martin Luther King, Jr. and Cesar Chavez helped lead movements to make better lives for millions of people.

> *Nothing great was ever achieved without enthusiasm.* —Ralph Waldo Emerson (1803-1882, essayist, poet, and leader of the transcendentalist movement)

Patience: The ability to wait without becoming annoyed or upset, and to be angry without losing one's temper.

Patience, although difficult, can be a time saver and a lifesaver. During a conflict, patience means having to cool down before dealing with the situation. Losing one's temper may cause impulsive reactions and may even endanger people. Working in groups can test a person's patience because it generally takes more time to get things done than if working alone. Leadership takes patience on many levels, from being a good listener to encouraging someone to do something you already know how to do.

Patience serves as a protection against wrongs as clothes do against cold. For if you put on more clothes as the cold increases, it will have no power to hurt you. So in like manner you must grow in patience when you meet with great wrongs, and they will be powerless to vex your mind.
—Leonardo Da Vinci (1452 - 1519, Italian polymath: scientist, mathematician, engineer, inventor, anatomist, painter, sculptor, architect, musician, and writer)

Perseverance: The ability to be steady and continue action despite difficulties or setbacks.
Perseverance and patience are related in that they both mean persistence in the face of difficulty. The difference is that perseverance tends to be more active, while patience is more passive. To persevere means to not give up. When frustration sets in, perseverance is one solution. Figure out how to deal with the frustration and get back on track.

Perseverance is the hard work you do after you get tired of doing the hard work you already did.
—Newt Gingrich (b. 1943, politician)

Resourcefulness: To solve problems creatively.
Jury-rigging is a form of resourcefulness. It means to use whatever is on hand to make what you need. If, for example, you need to hold something together for a short time and don't have tape or glue, you use the bubble gum you have been chewing. Being resourceful in problem solving means to think outside of the box and entertain creative and innovative ways of getting something done. Many groups do not have the money to do everything they want to do. Recruiting volunteers, using things more than once or for more than one purpose, and making due with what you have are ways to be resourceful.

The mantle of leadership falls on those who try to understand and abide by the rules (respectful), who keep trying, or try new ideas when there's a setback (resourcefulness) and who face up to the consequences of their actions (responsibility). —John E. Anderson (business and civic leader)

Sense of Humor: The ability to be playful and enjoy things that are fun or funny.
A good sense of humor can go far in relieving stress and making connections with others. Seeing the humor in a situation that is challenging or frustrating can help to break the tension and move things along. With a sense of humor, however, comes responsibility. Laughter can be used to hurt or heal. If used compassionately, humor can do wonders to make things go smoothly. If, on the other hand, people are laughed at, they may feel singled out, embarrassed, or excluded.

A sense of humor is part of the art of leadership, of getting along with people, of getting things done.
—Dwight D. Eisenhower (1890-1969, soldier, politician and president of the United States)

Service: The act of doing something for others.
Service is the ability to think about, and do things, for others. It requires action. It is more than charity, in that charity means to give something to others, while service means to do something for, and with, others. A true leader serves people rather than one's own self-interest.

We must be silent before we can listen.
We must listen before we can learn.
We must learn before we can prepare.
We must prepare before we can serve.
We must serve before we can lead.
—William Arthur Ward (1921-1994 author)

2.B sequence 6

Journal assignment.
Using the list of traits brainstormed by the class and the life skills that support leadership, write about the traits you feel are strong for you, and the ones you would like to develop more.

2.B sequence 7

Have students choose one life skill and write a SMART goal in their journal to work on that life skill during the next month (see p. ?? for a review of SMART goals).
Ask students to create a progress sheet to include in their portfolio.

2.B sequence 8

Journal assignment.
Write a personal definition of leadership. As more information and insights are gained about leadership, they will have an opportunity to modify this definition.

Lesson 2.C Overview: Compare and Contrast the Notions That Leaders are Born and/or Made

Focus:
- Seek consensus about group decisions
- Encourage diversity of opinion
- Practice reflection
- Examine the ethical framework in which you make decisions and interact with others
- Inventory your internal and external assets
- Seek diverse ideas and points of view

Step	Activity	Time	Assessment
1	Facilitate the Continuum activity to explore the gradations of either/or concepts.	15 - 20 min	
2	Foreshadow subsequent concepts.	5 - 10 min	
3	Watch videos or video clips of leaders who have been identified, born into the role, or who have grown into the role by circumstance.	1.5 - 2 hr	
4	Fill out video guide observation sheet.	20 - 30 min	assessment
5	Share thoughts about the films with the class.	10 - 15 min	
6	Share definitions of aristocracy and democracy.	10 - 20 min	
7	Share skits comparing aristocratic and democratic scenarios.	20 - 30 min	assessment
8	Read opinions on whether leaders are born or made.	30 - 45 min	
9	Hold a debate about whether leaders are born or made.	20 - 30 min	assessment
10	Review consensus guidelines.	5 - 10 min	
11	In base teams, attempt to reach consensus on whether leaders are born or made.	10 - 20 min	assessment
12	Journal assignment.	10 - 15 min	assessment

Time Frame: 4 - 6.5 hours

Lesson 2.C Sequence: Compare and Contrast the Notions That Leaders are Born and/or Made

Facilitate the Continuum activity to explore the gradations of either/or concepts.

activity

Continuum

Focus: Exploring gradations of either/or concepts
Materials: A line drawn on the board, list of either/or scenarios
Time: 15 - 20 minutes
Sequence: Deinhibitizer/Trust
Sources: See *Chiji Pocket Processing Cards*: Institute for Experiential Education (IEE)of La Crosse, WI, 608-784-0789. These Yin & Yang cards can be great tools for an activity such as this.

Suggested Procedure

1. Draw a horizontal line across the board to make a continuum.
2. Present one of the either/or scenarios listed below (or make up your own):
    - Saving money vs. spending money
    - Jumping into action vs. being patient
    - Morning person vs. night owl
    - Following the rules vs. doing your own thing
    - Doing something challenging vs. doing something easy
    - Competing vs. cooperating
    - Moving on vs. not giving up
    - Trying a new way vs. sticking with the reliable
    - Being serious vs. being silly
    - Optimistic vs. pessimistic
    - Being active vs. being sedentary
    - Goal oriented vs. process oriented
3. Identify one part of the scenario as one pole of the continuum, and the other as the other pole of the continuum.
4. Have students place themselves on the line according to what fits them best.
5. After students have placed themselves for a given scenario, ask clarifying questions or have a few students share why they put themselves where they did.

Sample Processing Questions

- Did you notice any patterns about where you placed yourself over the series of scenarios?
- Did we have a scenario where everyone put themselves only on the ends, or were there always a range of people across the continuum?
- Why are only either/or situations rare? Why do we generally have a range of possibilities?

Facilitation Notes

The purpose of this activity is to explore the idea that we rarely encounter simple either/or situations. As the class continues to look at dichotomies such as aristocracy/democracy and are leaders born or made, this lesson becomes important.

2.C sequence 2

Foreshadow subsequent concepts.

Note that the next activities present dichotomies, where things will be presented in an either/or format. Remind students that even though the following concepts are presented this way, the preceding activity illustrates the notion that few things are either/or in our world. Challenge your students to look for the subtleties between the either/or views.

Watch videos or video clips of leaders who have been identified, born into the role, or who have grown into the role by circumstance.

Watch one or more of the movies from each list below, and/or watch films you identify as ones that are appropriate to compare and contrast leadership through identification/birth and leadership by circumstance.

Leadership through identification: *Kundun* (Dalai Lama), *The Madness of King George*, *The Queen* (Queen Elizabeth)
Leadership by circumstance: *The Story of Rosa Parks*, *Norma Rae*, *Gandhi*

Foreshadow the films with reflection questions:

- What caused the people in the film to assume leadership positions – what got them there to begin with?
- Were there strong rules about how they should act or not act as leaders? In what way?
- How did their way of handling leadership change as time went on? What caused this?
- When did the person highlighted in the film decide that he or she was leading? What caused this?
- When did other people recognize them as taking a leadership role?
- Did how they come into a leadership role determine how they handled the role? If so, in what ways?

Fill out video guide observation sheet.

Using the reflection questions from above, work with one or two other people to answer them (students may also choose to work alone, if they wish).

Share thoughts about the films with the class.

Hold a discussion about the answers/questions from the video guide sheet. Ask students to share their observations about the way the characters in the films approached and handled their respective situations.

Discuss the question: What is the difference between the two situations?

Share definitions of "aristocracy" and "democracy."

Aristocracy

- A hereditary ruling class; nobility.
- Government by a ruling class.
- A state or country having this form of government.
- Government by the citizens deemed to be best qualified to lead.
- A state having such a government.
- A group or class considered superior to others.

Democracy

- Government by the people, exercised either directly or through elected representatives.
- A political or social unit that has such a government.
- The common people, considered as the primary source of political power.
- Majority rules.
- The principles of social equality and respect for the individual within a community (Answers.com, 2007).[3]

Each of these systems is loosely based on the idea that leaders are either born or made. Share some governmental examples of aristocracies and democracies today. Many countries now have both operating simultaneously (England, for example has both a monarch and parliament).

Make a connection between the ideas of aristocracy and democracy and how it might look in the lives of the students. For example, some of their peers may be acknowledged leaders by virtue of their academic or athletic abilities, or because of their family background. On the other hand, there are those who choose to be part of student government or make choices to be part of groups and take on leadership roles.

Acknowledge that we all have biases about whether one way of viewing leadership is better than the other, and that the class will have an opportunity to debate the matter soon.

Share skits comparing aristocratic and democratic scenarios.

Have students get into base teams, and give each team a scenario card. They are to create a skit showing how it might be handled from an aristocratic perspective and a democratic perspective.

Scenario card: You are having a party. Who do you invite?
Scenario card: Your group is ordering a pizza. How do you decide which pizza to get?
Scenario card: You want to go to college. Which school do you go to?
Scenario card: It's time for basketball try-outs. How do the try-outs go? How are decisions made about who makes the team?
Scenario card: The school is having a dance. How does the planning go?
Scenario card: You are at a lunch buffet. When and how do people get to eat?
Scenario card: Something is broken. How does it get fixed?

After each skit, ask the class to notice the differences and similarities between the two perspectives.

Read opinions on whether leaders are born or made.

Now that the students have looked at the dichotomy of aristocracy and democracy, introduce another dichotomy about the debate between leaders being born or made. Are people born with leadership qualities such as powerful personalities and charisma, can people learn leadership skills like courage and integrity, or does it lie somewhere in between?

Are Leaders Born or Made?

Aristocratic thinking assumes that people are entitled to leadership based on the family in which they were born. True democracy, on the other hand, grants everyone the right to participate in the system. In one view, a few people have the power based on who they are, while in the other, everyone can have power.

One of the long-standing debates about leadership is whether leaders are born or made. In an aristocracy, it is presumed that some are born into leadership; but does that, in itself, make someone a good leader? Even in a democracy, it can be argued that some people are born with personalities and talents that make them natural leaders, but can these skills be learned? Following are arguments for both sides of this debate. Like most dichotomies (opposites, or things that vary widely), there may be some middle ground. If we create a continuum, it would look like this:

Where do you fit on the continuum?

LEADERS ARE BORN

Great Leaders Are Born; Great Managers Are Made

As the title suggests, Ron Morris distinguishes between leaders and managers. In his view, true leaders are like an alpha wolf in the pack. When one gets into a group, everyone else submits to him/her. Some people may look like they have leadership skills, but they are not true leaders. He calls these people "interim leaders" because when a true leader arrives, they naturally follow that person.

Morris says that these born leaders are not necessarily the smartest, loudest, or biggest person in the group, but people just know that they are the leaders. They have a way about them, a charisma, that makes people trust them. The fact that even young children show these traits proves that it is not something that anyone taught them, but something they were born with. You can see this when someone just makes things happen. Sometimes this person asks for advice, and sometimes s/he just did it because it was easier to do it than to spend the time telling another how to do it.

Managers, on the other hand, "are made, and not born." Morris considers management unnatural and stressful, requiring the unnatural, but learnable skills such as confrontation, patience, and planning. It takes a whole lot of work to organize people to get things done.

In short, Morris believes that true leadership is easy because it comes naturally to those who were born that way. Management, on the other hand, is a whole lot of work. Both, though, are necessary to get things done, and it pays to know the difference between leadership and management. Mixing them can cause trouble.

Ron Morris clearly states the belief that leaders are born and not made: "To employ one of my favorite aphorisms, 'The art of leading others is factory installed, and that's just the way it is.' You either have leadership qualities, or you do not have leadership qualities. End of story."[4]

Leaders Are Made

The 108 Skills of Natural Born Leaders by Warren Blanck

The idea of born leaders is a myth. Blanck asserts that "No one is genetically programmed ... as a leader.... Some people are labeled natural born leaders because they effortlessly, spontaneously, consistently, and frequently demonstrate the specific skills that cause others to willingly follow" (Blanck, 2001).[5] And, people can learn these skills, if they pay attention to them.

Blanck proposes 108 skills that help people develop as leaders and divides them into nine categories:

1. **Self-Awareness:** This means knowing yourself, and to do that means paying attention. Learning from experience, you can be better at managing time, responding to stress and crisis, and be able to balance all the demands that you face everyday.
2. **Capacity to Develop Rapport with People:** Another way to say this is that you develop relationships with others based on trust, consistency, and integrity. You are friendly, sincere, and caring toward others and appreciate others frequently.
3. **Ability to Clarify Expectations:** This is about being clear and firm in what and how things get done. Creating clear ground rules, or keeping an eye on the shared vision and reminding people of it are examples of this skill set. Another example is not getting bogged down in hearsay and rumor mongering, but checking things out and getting the facts before reacting to situations.
4. **Ability to Map the Territory to Identify the Need to Lead:** To use this skill set means that you don't stay in an office and make decisions, but go out and get information from many points of view. When getting information, it is important to look at it on many levels—local, global, short term and long term. The learning never stops.
5. **Ability to Chart a Course of Leadership Action:** Rather than reacting to situations impulsively and without much thought, you have a plan of action to decide what needs to be taken on first and what can be handled later. When taking action, it is for a reason, not because there is nothing better to do.
6. **Ability to Develop Others as Leaders:** In this view of leadership, diversity is seen as a strength. It is not about doing things for people—it is about doing things with them, and allowing them to take things on. People can act as mentors to help people develop their skills.
7. **Ability to Build the Base to Gain Commitment:** It is important to know that you are not alone. Using these skills means that you consciously build alliances with both those inside and outside your group. Sharing power and supporting people is important.
8. **Ability to Influence Others to Willingly Follow:** This has as much to do with communication as it does with the power of persuasion. If people don't know why they are being asked to do something, it is difficult for them to follow your lead. Communicating your message, and meeting people where they are, rather than trying to force them, can make a huge difference.
9. **Ability to Create a Motivating Environment:** This means that people know what they are being asked to do, have clear roles, and know that what they are doing will make a difference. Conflicts are handled skillfully so that people feel heard and involved in win-win solutions.

Leading, in Blanck's view, means working hard at developing the right skills. It just takes the motivation to start and the commitment to keep at it.

Hold a debate about whether leaders are born or made.

Randomly divide the class in half and assign them a side: leaders are born or leaders are made. Give them time to prepare their arguments.

Ask each side to assign a spokesperson. Tell them that they will have a total of 6 minutes to present their arguments in this fashion:

- Opening statements: start with leaders are born, follow with leaders are made: 1 minute
- Leaders are made: 2 minutes
- Leaders are born: 2 minute rebuttal
- Caucus with their respective side: 3 minutes
- Leaders are born: 2 minutes
- Leaders are made: 2 minute rebuttal
- Closing statements: start with leaders are made, follow with leaders are born: 1 minute each

2.C sequence 10

Review consensus guidelines.

Guidelines for Reaching Consensus

A. Avoid arguing for your own ranking. Present your position clearly and logically. Listen to the other member's reactions and consider them carefully before you press your point.

B. Do not assume that someone must win and someone must lose when a discussion reaches a stalemate. Instead, look for acceptable alternatives for all members.

C. Do not change your mind to avoid conflict and reach agreement and harmony. Explore the reasons decisions are made and be sure everyone accepts the solution for basically similar or complementary reasons. Yield only to positions with logical and objectively sound foundations.

D. Avoid conflict-reducing techniques such as majority vote, averages, coin flips, and bargaining. When a dissenting member finally agrees, avoid feeling s/he must be rewarded with his/her own way at a later point.

E. Differences of opinion are natural and expected. Seek them out and try to involve everyone in the decision-making process. Differences of opinion can help group decision making by providing a wide range of information.

Fist to Five Strategy

This is a way to "vote" on a decision to see if a consensus exists.

A. State the decision that is to be made so that everyone understands.

B. Ask for a fist-to-five decision. This is where everyone puts out 0 to 5 fingers depending on how they feel about the decision:

    5 fingers = The best idea ever!
    4 fingers = It's a really good thing to do!
    3 fingers = It's okay
    2 fingers = I can live with it
    1 finger = I won't block it
    0 fingers = Block

C. Look around to assess the quality of the decision. If no one blocks it, then a decision has been made. If, however, even one person blocks it, then more negotiating must take place before moving on. If everyone puts out one or two fingers, then it might also behoove a group to continue working through the problem to reach a better decision.

2.C sequence 11

In base teams, attempt to reach consensus on whether leaders are born or made.

Each team attempts to reach consensus on whether or not leaders are born or made. They may or may not actually reach a consensus. Report the outcome to the class.

Journal assignment.

Revisit your working definition of leadership and modify it as necessary.

Lesson 2.D Overview: Explore Different Views of Leadership[6]

Focus: Practice reflection
Foster a climate of continuous learning
Foster creativity
Facilitate brainstorming

		assessment	
1 Facilitate the Leadership Egg Drop activity to explore traditional views of leadership. *1 - 1.5 hr*	**2** Hold a discussion using processing questions from the activity. *15 - 20 min*	**3** Present traditional views of leadership. *10 - 20 min*	**4** Students create examples of when each traditional view of leadership might be appropriate to use. *10 - 20 min*
	assessment		portfolio
5 Present the concept of Situational Leadership®. *15 - 20 min*	**6** Journal assignment. *10 - 15 min*	**7** Brainstorm a list of acknowledged leaders. *10 - 15 min*	**8** Research a project on one of the people who is an acknowledged leader. *varies*
portfolio		assessment	
9 Create a collage depicting their chosen leader's style(s). *30 - 60 min*	**10** Write a 100- to 200-word caption describing the collage. *30 - 45 min*	**11** Post collages and captions for review. *15 - 30 min*	

Time Frame: 4 - 6 hours

Lesson 2.D Sequence: Explore Different Views of Leadership

Facilitate the Leadership Egg Drop activity.
Facilitate the Leadership Egg Drop activity with different rules for different groups (autocratic, democratic, laissez-faire).

activity

Leadership Egg Drop

Focus: Integrity, cooperation, problem solving
Materials: A raw egg for each group of 4 to 5 (have a couple extra to allow for breakage), 25 straws for each group, 3 feet of masking tape for each group, garbage bag, large paper and markers, copies of the Leadership Stucture sheets.
Time: 1 - 1.5 hours
Sequence: Problem Solving
Sources: Unknown

Suggested Procedure

1. Divide the class into small groups of 4 to 5.
2. Give each group the Leadership Structure sheet (p. 188) that applies to them. The sheets contain information about what they are to do and how they are to do it.
3. **Your role as the "boss" or "manager"** is different with each group.

For the **Autocratic groups**: Give them the sheet of paper with explicit instructions (only give them the instructions they need for whatever they are working on at the time).

Instructions for making the device:
Using all but 2 straws, tape them together into a raft. This will be the landing pad for your Integrity Egg. Wrap the other two straws around the egg, one the short way and the other the long way, and tape together. Draw a happy face on the egg.

Instructions for naming the device:
You will name your device the "In-tegg-rator."

Instructions for creating the ad:
Draw a picture of the In-tegg-rator flying through the air toward a wall that has the words: "Peer pressure," "Shop-lifting," "Lying" written on it. Hold this up to the group and assign someone to say: "In life there are many ways to have your integrity tested. The In-tegg-rator is your first layer of protection! Call now for a free brochure. 888-555-1234."

For the **Democratic groups:** When they check in with you, simply ask a few questions and see if they are happy with what they have designed. Then tell them that everything looks good and to continue with the fine work they are doing.

For the **Laissez Faire groups:** You only answer their questions. If they do not think to ask certain questions, do not divulge any extra information.

4. Give teams 15 minutes to work on their device, and another 10 minutes to create their advertisement (they can choose to do these simultaneously by having their group divide the work).
5. When all teams are ready, have each group show their advertisement to the large group and then test their device by dropping the egg from a height of 8 feet. They should choose someone from their group to drop the egg. **This is the time to use the garbage bag as a landing spot for the egg...**
6. Encourage the wonderful use of puns by saying things like, "That's eggs-actly what I meant," or "Isn't this egg-citing?!" When you're done, you can all act like eggs and scramble....

Sample Processing Questions

- How did you work together to accomplish the tasks?
- What influenced the way you did your task?
- How did you like the leadership structure that was assigned to you? What worked and didn't work for you?
- How did it make you think about the meaning of integrity? What insights did you gain about what it means and how it works?
- What situations are out there that test your integrity?

Facilitation Notes

Be prepared for some frustration from individuals and groups. The Autocratic groups may be very happy with the structure or constrained by it, especially since some will look at the idea and think, "this won't work." The democratic groups may get tired of checking in with you, but may like the immediate feedback. Finally, the Laissez Faire groups may love to just do whatever they think should be done, observing other groups to get ideas, or going crazy with the lack of direction. It all depends.

This activity can get quite competitive, so it may be necessary to set some ground rules before the eggs are dropped. It's a nice time to make connections about how, when our integrity is tested, we sometimes crack. These can then be learning situations.

Leadership Structure Sheets

Leadership Structure Sheet 1: Autocratic

You will need a raw egg, 3 feet of masking tape, and 25 straws, which the teacher will give you when s/he is ready. The egg represents your integrity. It is fragile, and it is about to be tested.

Your task is to create protection for the Integrity Egg to remain strong and not falter or break when faced with a difficult decision. You are to do this using only the materials that are given to you.

The teacher will tell you how s/he wants you to do this. You must follow directions to the letter. Make sure you do not do anything to the egg until s/he has told you it's okay. If you are not sure what to do, ask questions.

Your second task is to name your device. When you are ready, ask the teacher, and she will tell you what to call it.

Your third task is to create an advertisement for your protection device that shows us how it works. You will have paper and markers to use for this. When you are ready for the materials, ask the teacher. She will then tell you how this is to be done. Make sure to follow her directions.

Leadership Structure Sheet 2: Democratic

You will need a raw egg, 3 feet of masking tape and 25 straws, which the teacher will give you when s/he is ready. The egg represents your integrity. It is fragile, and it is about to be tested.

Your task is to create protection for the Integrity Egg to remain strong and not falter or break when faced with a difficult decision. You are to do this using only the materials that are given to you. When finished, you will have the opportunity to drop the Integrity Egg from 8 feet to see how well your protection device works.

You may discuss how you want to do this. Before starting to build the device, check with the teacher to see if you're on the right track. S/he will listen to your ideas and discuss any possible problems. Once you have checked in, create your device.

Your second task is to name your device. Again, check with the teacher before settling on a final name.

Your third task is to create an advertisement for your protection device that shows how it works. You will have paper and markers to use for this, if you wish. The materials are available from the teacher. Before putting pen to paper, develop your ideas and check in with the teacher to make sure you are on the right track.

Leadership Structure Sheet 3: Laissez Faire

Your materials are: One egg, 25 straws and 3 feet of masking tape. You can also use markers and large paper. If you have questions, ask the teacher. Do not ask questions of other groups.

Hold a discussion using processing questions from the activity.

Take the opportunity to compare the different structures. Have students describe what they went through and what they think about each of them.

Present traditional views of leadership.

Traditional Views of Leadership: Autocratic, Democratic, Laissez Faire

In 1939 a significant leadership study by Lewin, Lippitt and White[7] identified leadership styles at a summer camp with boys. The three major styles identified were Autocratic, Democratic, and Laissez Faire. On the following page is a comparison of the three.[8]

Using definitions of the traditional views of leadership, have students create examples of when each might be appropriate to use.

Either alone or in pairs, ask students to think of an example when each traditional leadership style might be useful. For example:

- **Autocratic:** During an emergency, or during a game when the coach needs to make a decision on strategy.
- **Democratic:** When there are two or more choices to make, or when a group is having difficulty arriving at a decision.
- **Laissez-Faire:** When participants really know their job, or when the stakes are low.

Ask students why they wouldn't use just one style of leadership. Why mix and match styles? This will lead to the concept of Situational Leadership.®

TRADITIONAL VIEWS OF LEADERSHIP

In 1939 a significant leadership study by Lewin, Lippitt and White identified leadership styles. They identified three major styles: Autocratic, Democratic, Laissez Faire. Following is a comparison of the three.

Leadership Style	Autocratic	Democratic	Laissez-Faire
Definition	Also known as "authoritarian leadership," a person using this style makes decisions and then passes them on to others.	Also known as "participative leadership," there is equal participation in the decision making processes of an organization or group.	Also known as "delegative leadership," this is the practice of letting people do as they wish.
Description	An autocratic leader tells people what to do and how to do it—even down to the smallest details. The leader makes all of the decisions and holds all power. This person gives out praise and criticism of others' work as needed.	A democratic leader works with the group to create policies. S/he gives an overview of the work to be done and allows group members to decide how best to do it. Group participation is encourages but the final decision is the leaders.	A Laissez-Faire leader is hands-off. A group is given complete freedom and no feedback unless they ask for it. Group members are the decision-makers.
Findings: Satisfaction	How the leader used power was important. If they were bossy, people were more put out than if the leader held high standards.	People tended to feel satisfied with making their own decisions.	Members can lack any feeling of accomplishment because they don't know what is expected of them.
Findings: Productivity	Productivity was affected negatively if the group was under constant pressure by the leader, but was high when people felt their efforts showed progress. Work slacked off, though, when the leader was not there.	Group members tended to listen to each other and were more open to ideas. Self confidence grew as the group achieved its goals.	A general lack of direction can cause people to be confused about whether they are accomplishing anything.
Findings: Relationship to the Leader	People generally either love or hate this leader.	Leader and group members are more like peers.	Not much of a connection with the leader.
Findings: Conflict	People can become dependent on the leader, enjoy the strong structure, and, if they accept the structure, tension between group members can remain low.	The pressures from the group can cause groupthink and a feeling that one must conform to the group.	The group does not come together. Competition can develop as people try to figure out what to do.
Best used when...	This style of leadership works best when time is limited and/or when the person seen as the leader has the most knowledge/skill.	This style of leadership works best when there is time to hold discussions and there is a variety of skill and knowledge to be shared.	This style of leadership works best when group members are highly skilled, knowledgeable, and motivated.

SITUATIONAL LEADERSHIP[9]

Not all situations call for a similar approach. This is the main idea behind Situational Leadership. This view of leadership takes into consideration that every group of people in a given time and place has different needs in order to work well together. It is the leader's responsibility to pay attention to the situation and provide the type of leadership that will work best for this group at this time in this place.

Situational Leadership takes into account how much experience, knowledge, skill, and maturation the people have in a group compared to the task. Some people simply do not want the responsibility of making decisions, so appreciate a hands-on (even autocratic) leadership style. Others want to be given a task and allowed to run with it. This type of leadership also relates to groups. When dealing with brand new and experienced people a leader gives more direction. On the other hand, if a group of people have been working together for a long time, have a lot of skills, and understand what they need to do, they can be more independent.

Hersey and Blanchard present four stages of leadership that fit particular situations[11]:

Stage 1: Directing or "telling": Low responsibility and low independence. People need to be told what to do and how to do it. Feedback should be frequent so they know they are accomplishing their assignment correctly.

Stage 2: Coaching or "selling": Low responsibility and medium independence. People need to be involved in creating ideas for how to get their task completed, but don't have the whole picture. This means that they cannot make decisions on how to change things on their own. They still need lots of feedback so they can grow.

Stage 3: Supporting or "participating": Medium responsibility and medium independence. People have confidence in themselves to share in decision making, but still need support from the person in authority. If left alone, they will feel insecure and fall back to the coaching stage.

Stage 4: Delegating: High responsibility and high independence. People need only to be told what the expectations are and then they go out and do it. Although they do not need much feedback, it is still a good idea to touch base now and then to make sure they are on the right track.[10]

Situational Leadership® requires that the person in charge know the people with whom they work. They must pay attention to whether the group members have the skills to get things done, and also if they are interested in doing the work. The more independent and responsible individuals and groups are, the higher the stage they can operate. Problems arise when leaders make these decisions based on how they like to be led, rather than on how the people in their group like to be led. These are two very different perspectives. So the next time you find yourself in a position of leadership, whether as an authority or as a member of a group, remember that not all situations are equal.

2.D sequence 6

Journal assignment.

Revisit your personal working definition of leadership. Modify as necessary.

2.D sequence 7

Brainstorm a list of acknowledged local, national, and international leaders.

As a class, come up with a list of acknowledged leaders: political leaders, corporate leaders, community leaders, artistic leaders, etc.

2.D sequence 8

Research project.

Each student researches the leadership styles of one person from the list as they relate to autocratic, democratic, laissez faire, and situational styles of leadership.

2.D sequence 9

Create a collage depicting their chosen leaders' style(s).

Using pictures from magazines, their own drawings, etc., have students create a collage that shows the leader they chose to research and how s/he uses the traditional styles of leadership and/or Situational Leadership.

2.D sequence 10

Write a 100- to 200-word caption describing the collage.

Students write a caption that describes their collage and how it relates to their chose leader and that leader's styles.

2.D sequence 11

Post collages and captions in the room for review.

After viewing, have students add the collage to their portfolios.

Lesson 2.E Overview: Analyze the Roles of Risk Taking and Long-Term Vision When Leading

Focus:
- Develop capacity to act in the face of risk and uncertainty
- Develop capacity for long-term vision
- Listen actively
- Support risk taking

1	2	3	4
Facilitate the Growth Circles activity to explore the concept of risk taking.	Hold a discussion using processing questions from the activity.	Journal assignment.	Facilitate the Lines of Vision activity to explore the idea of creating a vision.
20 - 45 min	*10 - 15 min*	*10 - 15 min*	*20 - 30 min*

assessment

5	6	7	8
Discuss the experience using the processing questions from the activity.	Explain the difference between vision and mission statements.	Create a personal vision and mission statement.	In small groups, share your personal vision and mission statements.
10 - 15 min	*15 - 20 min*	*1 - 2 hr*	*15 - 30 min*

Time frame: 3 - 5 hours

LESSON 2.E SEQUENCE: ANALYZE THE ROLES OF RISK TAKING AND LONG-TERM VISION WHEN LEADING

Facilitate the Growth Circles activity to explore the concept of risk taking and leadership.

activity

GROWTH CIRCLES

Focus: Challenge by Choice, perspective taking, risk taking
Materials: Ropes or tape on the floor for making three concentric circles (see below)
Time: 20 - 45 minutes
Sequence: Trust
Sources: See *Adventure Education for the Classroom Community* by Frank & Panico, and *Journey Toward the Caring Classroom* by Frank.

Growth Circles look like this:

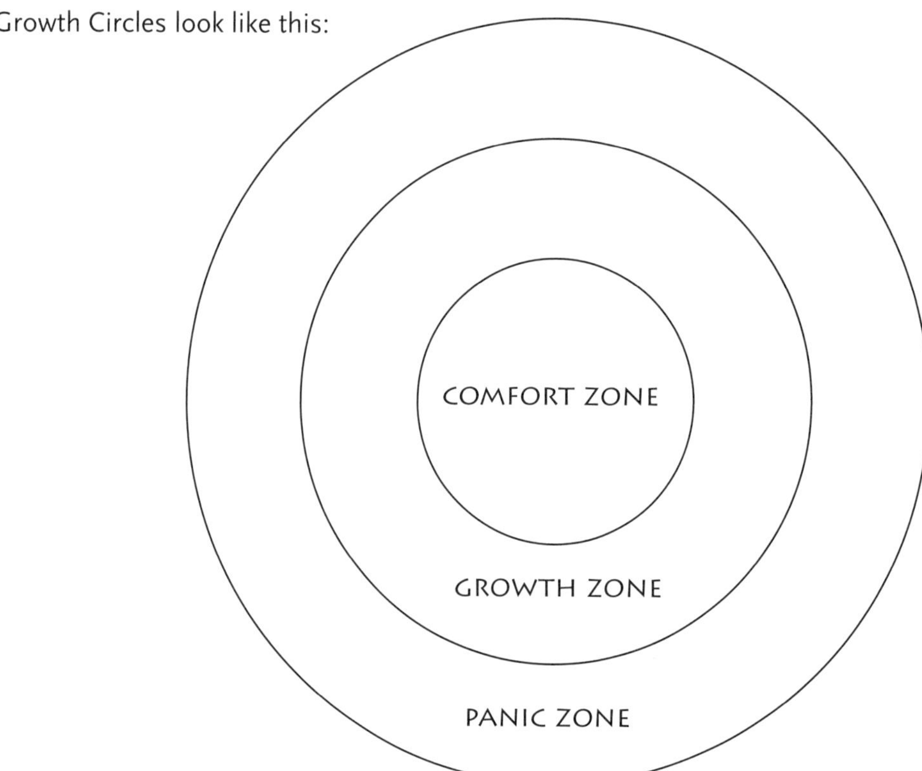

Suggested Procedure

1. Discuss the meaning of the growth circles (see below).
2. Outline the growth circles on the floor.
3. Ask questions, and have people put themselves where they feel it is most appropriate. Sample questions include the following. How do you feel about:
    - spiders?
    - speaking in front of a large group?
    - singing solo in front of a large group?

- singing in a choir?
- bungee jumping?
- telling a family member that you love him or her?
- heights?
- going white water rafting?
- confronting a friend about something he or she did or said?
- snakes?
- taking a math test?
- introducing yourself to someone new?
- taking a driver's test?
- walking in the woods alone at night?
- walking in the city alone at night?
- going to a party alone, where you don't know anyone?
- standing up for someone who is being harassed or bullied?
- making a choice that is different than all of your friends?
- telling people what you think

4. After each question is asked and people have moved into position, give the students a chance to comment on why they put themselves in that particular spot. Is there a story to share? Who is sharing their point-of-view?
5. After a few of your questions, ask the students to throw out any questions they have for the class.

Sample Processing Questions

- Were you surprised at where you ended up in comparison to others?
- How can we support the choices each of us makes?
- What does risk taking have to do with leadership?
- How can we encourage you to step into your growth zone without putting too much pressure on you? What kind of encouragement is useful to you?

Facilitation Notes

Leading is an act of risk taking. To choose to speak up risks that others may disagree, to offer suggestions or ideas is to risk having these ideas rejected, or even to sit back and allow others to do something that you have done before is to risk that it may not work (or work the way one wants it to work...). This activity helps the class to explore these issues. It also allows students to recognize that people have different limits, different comfort levels, and different ways of approaching situations. Through discussion, students can explore how to support each individual to stretch him/herself, and take on leadership roles when called for.

Discuss activity using the processing questions from the activity.

Explore the idea of risk-taking as an act of leadership. What are some examples of risk taking when assuming leadership roles?

Journal assignment.

How might risk-taking relate to you when in a leadership role?

Facilitate the Lines of Vision activity to explore the idea of creating a vision.

activity

Lines of Vision

Focus: Visioning, communication
Materials: Two different communication sheets, paper and pens for each student
Time: 20 - 30 minutes
Sequence: Problem Solving
Sources: See "Say What?" in *Silver Bullets* by Rohnke.

Suggested Procedure

1. Have students pair up and sit back-to-back with their partners.
2. Ask them to choose who will be the communicator, and who will be the receiver.
3. Give each communicator the first sheet of random lines and shapes. **Tell them not to show it to anyone.**
4. Give each receiver a paper and pen.
5. The task is for the communicator to describe what is on his/her paper, while the receiver attempts to draw what is being communicated. The receiver may ask questions of the communicator. Before beginning, tell them they will have 3 minutes to work on the drawing.
6. At 3 minutes, ask the partners to compare the receiver's drawing with the communicators' paper.
7. Now ask the students to switch roles (communicators become the receivers, and vice versa).
8. Give the new communicators the other sheet with the drawings of the house and sun. The same rules apply—**do not show anyone the drawing**. Give them 3 minutes to verbally communicate what is on the paper to their receiver who attempts to draw it.
9. At 3 minutes, ask them to compare what was drawn with the sheet that was given out.
10. Ask them to discuss with their partners if there was any difference between the two attempts.

Sample Processing Questions

- What did you decide about the two attempts? Were they the same or different?
- Both drawings used the same lines and shapes, only in one of them the lines and shapes made a recognizable picture. Do you think that made a difference?
- The picture can be described as a "vision." What do you think a "vision" is?
- What are some examples of a "vision"?
- The random lines and shapes can be seen as different tasks or ideas that can make the vision a reality. What are some examples of day-to-day tasks that help us achieve a vision?

Facilitation Notes

Invariably, people do better on the attempt with the picture because they can formulate a picture (or vision) of what they are trying to create. When someone says "draw a house with a square, a triangle for the roof, etc.," it is easier to know what they are being asked to do compared with "draw a square, now draw a triangle."

During the first attempt, many people will use up all of the time, while on the second attempt, many will get done early. Make a note of that so you can have a discussion about how much quicker people were able to get it done. See if people have theories on why that is. They generally will say it's because we had practice or a picture of what we were trying to draw. You can use this as the bridge to talk about vision and mission, which happens next.

Communication Sheets

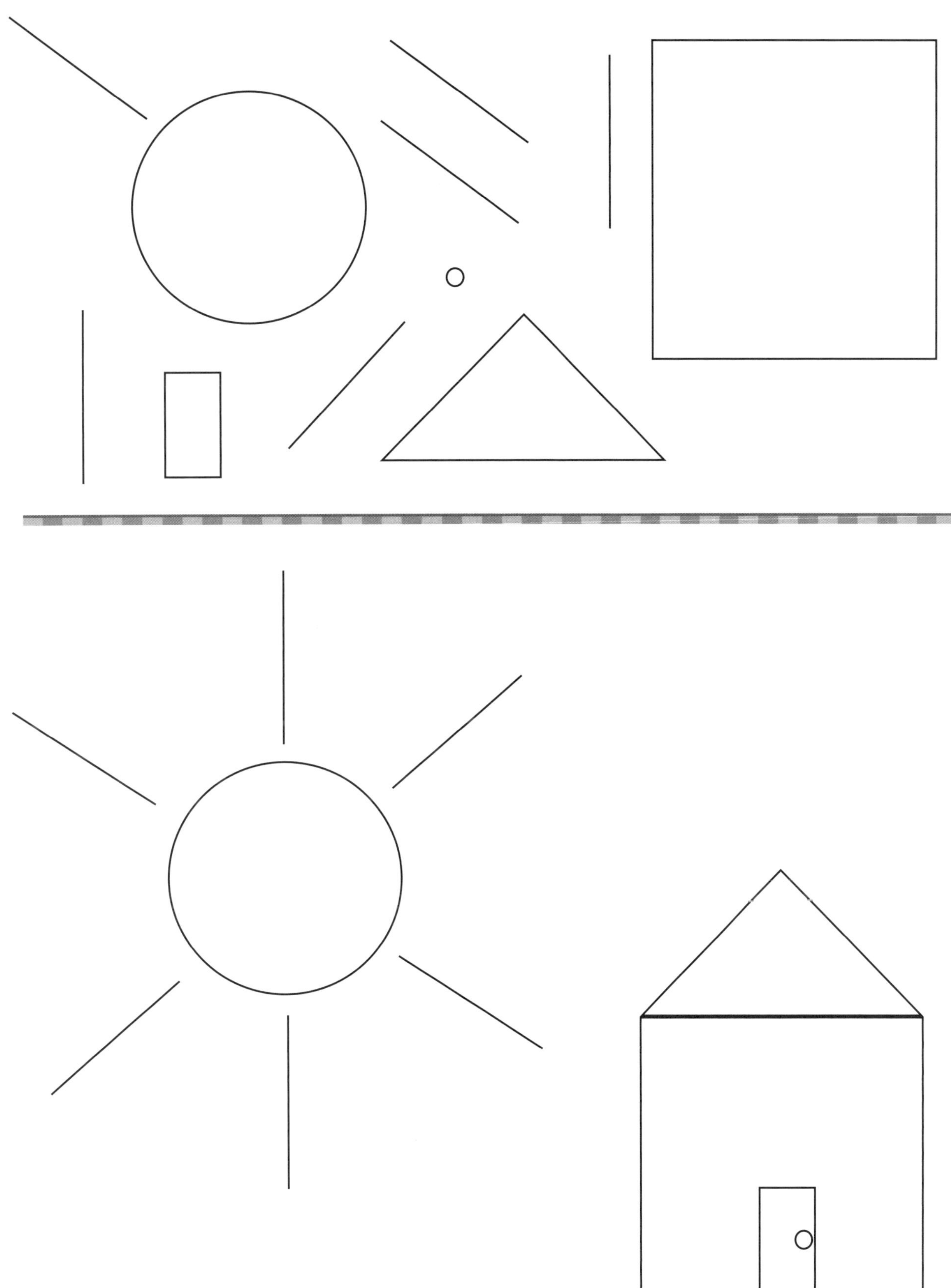

Discuss the experience using the processing questions from the activity.

Explain the difference between vision and mission statements to students. Include why organizations have them, and how they help a group focus. Discuss how people can also have personal vision and mission statements.

A vision is about the future—what you hope to see happen because of the work you are doing. In other words, how will your organization affect the world? A mission is the reason you exist. It is about what you do today to make your vision a reality tomorrow.

Vision Statements

A vision statement is a written statement of what the organization should look like in a certain number of years. This Vision Statement is the "guiding light" that provides the organization with an image toward which to strive.... The Vision Statement should be based upon the collective values and beliefs of the members of the organization.... The Vision Statement drives the goals of the organization." —**Jerry Valentine, Middle Level Leadership Center, 1998.**[11]

"A vision statement is a stated goal that provides direction, aligns key players and energizes people to achieve a common purpose. It states your organizational ideal, stretching the imagination and motivating people to rethink what is possible." —**Liberty Systems, 2002**

"Every organization has a destiny: a deep purpose that expresses the organization's reason for existence... To become more aware of an organization's vision, one must ask the members and learn to listen for their answers." —**Peter Senge**[12]

Good Vision Statements

- Are short, clear and to the point.
- Have meaning to those in the organization.
- Speak to the greater good.
- Inspire people.
- Allow people to see the big picture and think long-term.

Mission Statements

A mission statement is a written statement of the organization's purpose.... In other words, "What is the purpose of the organization within a larger context such as society?" —**Jerry Valentine, Middle Level Leadership Center, 1998.**[13]

"At its most basic, the mission statement describes the overall purpose of the organization." —**Carter McNamara, PhD**[14]

"A mission defines the area of business chosen to help make the vision a reality. It puts boundaries around the organization to channel and focus its efforts." —**Unknown Author**[15]

Good Mission Statements

- Are short and easy to remember.
- Are easy to understand and easily communicated.
- Explain what your organization is about and why it exists.

Examples of Vision and Mission Statements

Youth Tutoring Program

YTP Vision: All YTP students will be empowered to succeed academically and will possess the skills they need to realize their potential and achieve their hopes and dreams.

YTP's Mission is to tutor, guide and inspire at-risk youth living in public housing to achieve academic success. We create a challenging, safe and enriching environment where youth are matched with adults who offer academic support and mentoring throughout their school years. We partner with parents and advocate for students in their schools and in their communities.[16]

National Youth Leadership Forum

Vision: Established to help prepare extraordinary young people for their professional careers.

Mission: To bring various professions to life, empowering outstanding young people with confidence to make well-informed career choices.[17]

2.E sequence 7

Create a personal vision and mission statement.

Have students fill out the form on the following page that asks them pertinent questions. Then ask them to use the information to create a personal vision and mission statement. Write these in their journals. Add their personal vision and mission to their portfolios.

In small groups, share your personal vision and mission statements.

Split into small groups, and ask students to share their personal vision and mission statements, if they wish. Remind them that they have the right to pass, which means they can share any or all of their statements. They may also choose to not share any of it with the small group.

Creating a Personal Vision & Mission Statement

Please thoughtfully answer the following questions to help you reflect on areas that help you to formulate a personal vision and mission statement.[18] Write your answers in your journal.

1. If you were asked to create a TV show or write a book about something that you care deeply about, what would it be?

2. If you started a business or organization to solve a need, what would it be?

3. What subjects would you like to learn more about?

4. If you inherited a million dollars, and were required to give it away for a good cause, what would you do with the money?

5. People say, "Oh, you are so good at ____________."

6. Write down a list of at least 10 talents you have.

7. What is exciting about the world?

8. What angers or saddens you in or about the world?

9. If you had only six healthy months left to live, what would you do?

10. Imagine it is 10 years from now. Describe where you are, what you are doing, what you are wearing, and who you are with.

11. Imagine that you are middle aged (40 to 50 years old) and sitting with a friend, who asks, "What makes you happy about your life? What are you proud of?" What would you say?

12. Imagine that you have lived your whole life. Three things have happened for the better because of you during your life. What are these three things?

[18] The Franklin Covey Web site has a fun tool called the "mission builder." Check it out at http://www.franklincovey.com/fc/library_and_resources/mission_statement_builder.

Lesson 2.F Overview: Construct a Vision of an Ideal Leader

Focus: Encourage participation
Practice reflection
Develop capacity to act in the face of risk and uncertainty
Foster creativity

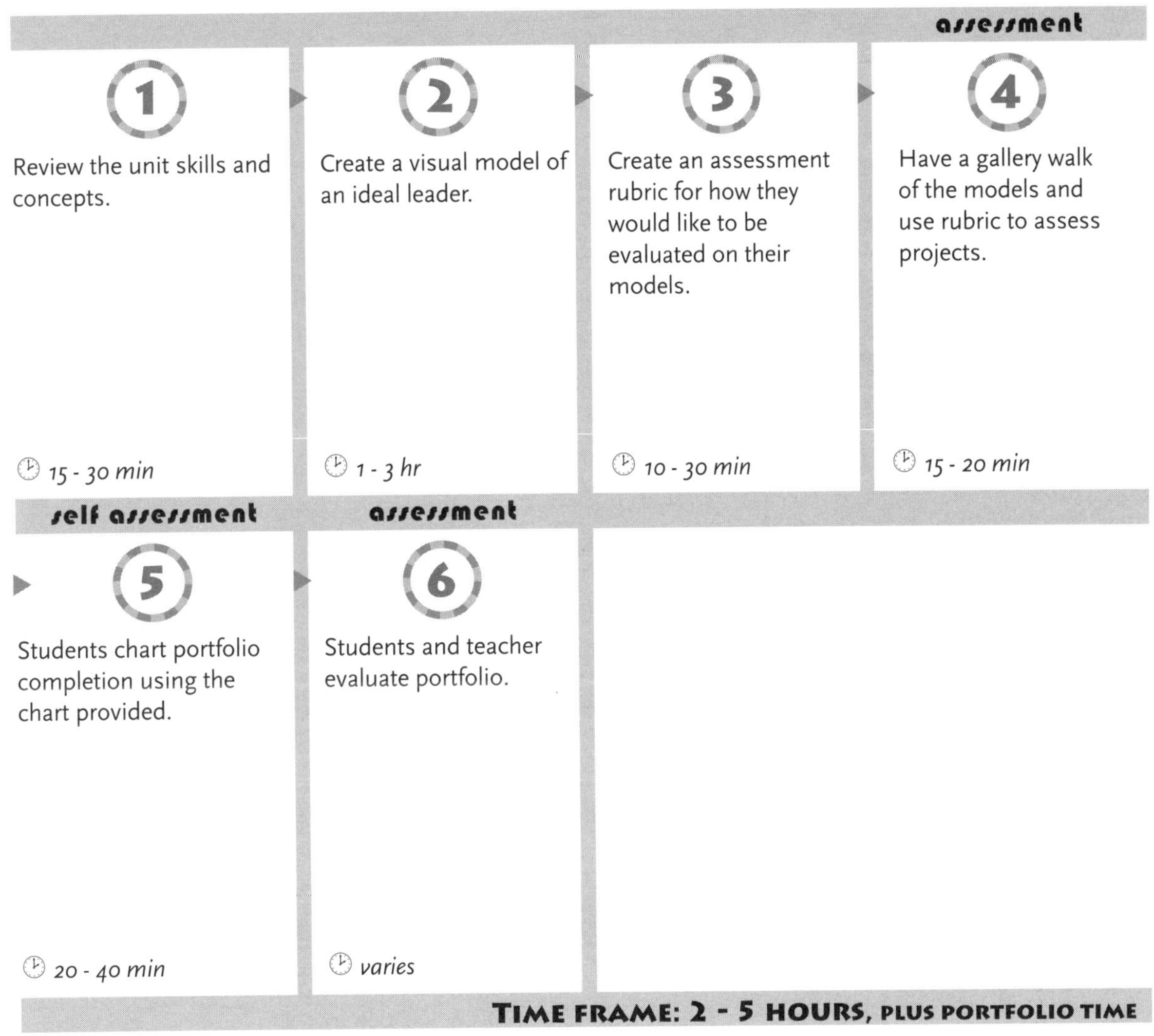

LESSON 2.F SEQUENCE: CONSTRUCT A VISION OF AN IDEAL LEADER

2.F sequence 1

Review the unit skills and themes. Check to see if there are any questions or gaps.

What is Leadership? Skills/Themes	Notes
Creating Community • Types of activities • Sequencing activities • Facilitating activities	
Personal Definition of Leadership • Leader role model • Qualities of a leader • Compose a working definition of leadership	
Are Leaders Born or Made? • Aristocracy and Democracy • Debate: are leaders born or made?	
Views of Leadership • Autocratic, Democratic, Laissez Faire • Situational Leadership® • Collage on an admired leader	
Risk Taking and Vision • Risk Taking • Vision and organization • Personal vision and mission	

Individual project: Create a visual model of an ideal leader.
Using everything saved in the portfolio that is applicable, show your model of an ideal leader in a visual way, e.g., visual arts, Power Point slide show, paper, statue.

Create an assessment rubric for how they would like to be evaluated on their models.
What criteria are important to show quality work? Have students think about the quality of the work and how it is presented. It may look something like this:

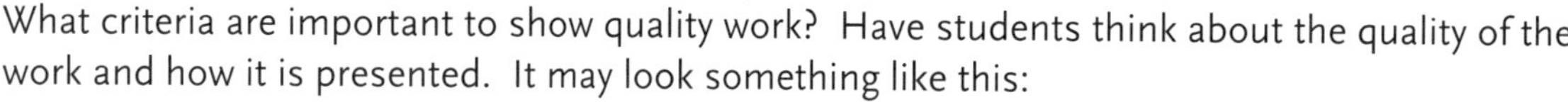

Criteria	1	2	3
Effort	Looks like it was put together at the last minute.	Obviously put some time into it.	Looks like a lot of time was spent working on this.
Quality	Lots of typos, mistakes, etc.	Clean and orderly.	A polished product.
Presentation	Lost interest and moved on.	Interesting.	Caught our attention.
Creativity	Wrote a paper.	Had pictures, photos, art work to go with the writing.	Created a multi-sensory product.
Etc.			

Model Evaluation Rubric

Criteria	1	2	3

2.F sequence 4

Have a gallery walk of the models and use rubric to assess projects.

Have a gallery walk of the presentations with self, peer, and teacher assessments using the agreed-upon rubric.

2.F sequence 5

Students chart portfolio completion using the chart provided.

Using the Portfolio Completion Chart on the following page, have students determine what assignments are in their portfolio for this unit, and reflect back on them by adding comments.

The purpose of this exercise is to help students remember what they have done and reflect on their learning. This is a list of assignments and assessments throughout the unit. There can certainly be more items in the portfolio than are listed here, especially if a student wishes to use it to keep track of his or her work.

2.F sequence 6

Students and teacher evaluate portfolio.

Evaluation will depend on the purpose of the class and the culture in the school. One way to evaluate portfolios is to have each student convene a group of 3 peers. The student then chooses a sampling of his/her work to share with this review committee. Their charge is to look over the work and then meet with the student to ask questions and provide feedback about the quality of the work. Before doing this you may wish to create some ground rules about how feedback will be offered. Using your established social contract is an option, as well.

The same sampling of work can then be evaluated by the student him/herself and the teacher. If necessary, grades can be suggested by each of the parties, and a grade can be assessed for the portfolio.

Portfolio Completion Chart

Item (Lesson)	Comments
Ice breaker, deinhibitizer, trust, and problem solving activities. (2.A)	
Facilitation self assessment using the provided rubric. (2.A)	
Paper about a person admired as a leadership role model. (2.B)	
Journal: Using the list of traits brainstormed by the class and the life skills that support leadership, write about the traits you feel are strong for you, and the ones you would like to develop more. (2.B)	
SMART Goal about one life skill to work on. (2.B)	
Journal: Write a personal definition of leadership. (2.B)	
Video guide about aristocracy and democracy. (2.C)	
Journal Assignment: Revisit your working definition of leadership and modify it as necessary. (2.C, 2.D)	

Portfolio Completion Chart (cont.)

Item (Lesson)	Comments
Collage and caption depicting a leader using traditional and/or Situational Leadership.® (2.D)	
Journal: How might risk-taking relate to you when in a leadership role? (2.E)	
Personal vision and mission statement. (2.E)	
Visual model of an ideal leader. (2.F)	

[1] For more on sequencing, see *Journey Toward the Caring Classroom*, pp. 23-27.

[2] A wonderful resource is *What Do You Stand For?: A Kid's Guide to Building Character*, by Barbara Lewis. See also *Building Everyday Leadership in All Teens* by Miriam MacGregor.

[3] Refer to: www.answers.com. Retrieved April 10, 2007.

[4] Refer to: *Great Leaders Are Born; Great Managers Are Made* (2003). Editor's Aide, Inc. Retrieved from: http://www.imakenews.com/techyvent/e_article000118147.cfm on February 8, 2005. Used with permission.

[5] Blanck, W. (2001). *The 108 Skills of Natural Born Leaders.* New York: AMACOM, American Management Association.

[6] Traditional and situational styles of leadership are discussed here. Another styles' comparison can be found in *Everyday Leadership*, where Miriam MacGregor compares masculine (traditional) and feminine (modern) styles of leadership (see bibliography).

[7] Lewin, K., Lippitt, R., & White, R. K. (1939). Patterns of Aggressive Behavior in Experimentally Created Social Climates. *Journal of Social Psychology, 10* 271-301.

[8] The creation of this chart was also informed by Van Wagner, K. (2007). *Lewin's Leadership Styles*. About.com: Psychology. Retrieved May 25, 2007, from http://psychology.about.com/od/leadership/a/leadstyles.htm

[9] Situational Leadership® is a registered trademark of the Center for Leadership Studies.

[10] For more information on Situational Leadership, visit Hersey and Blanchard's website: Center for Leadership Studies at http://www.situational.com/

[11] See Middle Level Leadership Center (MLLC) at http://www.mllc.org/index.php.

[12] See Change Facilitation. Retrieved from: http://www.change-management-toolbook.com/tools/vision.html

[13] See Middle Level Leadership Center (MLLC) at http://www.mllc.org/index.php.

[14] See Authenticity Consulting at http://www.authenticityconsulting.com/npod/about_us.htm

[15] See the Texas Youth Commission website at http://www.tyc.state.tx.us/archive/programs/vol_newsletter/fall_oct2002/oct2002_1.html

[16] See Youth Tutoring Program website at http://www.ytpseattle.org/about/mission.html

[17] See the National Youth Leadership Forum website at http://www.nylf.org/overview.cfm

[18] The Franklin Covey Web site has a fun tool called the "mission builder." Check it out at http://www.franklincovey.com/fc/library_and_resources/mission_statement_builder.

"One thing I know: the only ones among you who will be really happy are those who will have sought and found how to serve."

—Albert Schweitzer

UNIT 3 outline

What is Collaborative Leadership?

Collaborative leadership is different from traditional and situational types of leadership because it starts from the point of the group and not the individual. A common perspective on leadership places it within the individual. Collaborative leadership views the individual in the context of a group, hence the idea that leadership is built upon the foundation of relationships. Personal growth is a large part of the process, but it is not the only part.

This unit will bring the concepts of collaboration and leadership together. The students' explorations of both collaboration and leadership will put them in good position to make sense of the whole picture and put collaborative leadership into practice. It begins, again, with a deeper examination of relationships, now through the lens of sustaining community. Obviously, not all groups will come together with an ability or intent to do activities that help build and nurture relationships. Activities are but one tool that has proven highly effective in building bridges between people. This time, students will brainstorm the myriad of ways relationships can be created and then go out and try one.

From this base, we look at the wisdom and experience of two groups that have a rich experience with collaborative leadership: The core values of the Wisconsin Leadership Institute and the Qualities of a Collaborative Leader from Camp Manito-wish YMCA. Bringing these ideas together will help students focus on what collaborative leadership is. Then the class will put collaborative leadership into practice by taking on a service project to benefit the school or community.

The end of this unit is, of course, not the end. For these students, it is the beginning of their leadership journey. Now that it has begun, they will have no choice but to heed the call of being a leader among leaders.

A. Enduring Understandings

- Being a collaborative leader requires that one examines his/her personal values.
- Qualities of a collaborative leader mean different things to different people.
- There is a social component to being a collaborative leader that involves giving to the larger community.

B. Essential Questions

- How does one lead collaboratively?
- How is collaborative leadership similar and/or different from traditional and situational views of leadership?
- What does collaborative leadership mean to me?

C. Key Knowledge and Skills

Participants will know:

- The WLI core values of a collaborative leader.

- The Manito-wish 7 qualities of a collaborative leader.
- That leadership learning is a never-ending process.

Participants will be able to:
- Co-facilitate an activity with a large group.
- Self-assess collaborative leadership skills and set appropriate goals for personal development.
- Collaboratively plan and carry-out a service project.

D. Unit Activities/Performance Tasks: (Experiential Lessons)
- Continue to intentionally build and maintain a safe working environment.
- Assess personal collaborative leadership core values.
- Use collaborative leadership qualities and core values.
- Practice effective use of collaborative leadership qualities and core values.

E. Varied Classroom Assessments and Rubrics
- Facilitate a community/relationship building strategy with the class.
- Journal assignments.
- Revisit and revise paper on an admired leader to assimilate new information.
- Plan and carry out a small class gathering.
- Action plan on how to be a collaborative leader in one's own life.
- Portfolio evaluation.
- Course evaluation.

F. Student Self-Assessment
- Collaborative qualities/skills assessment.
- Write a letter to yourself about your experience with and your goals for collaborative leadership.
- Portfolio completion rubric.
- Portfolio evaluation.

G. Peer Feedback
- Debrief and evaluate a small group service project.
- Activity feedback form.
- Portfolio evaluation.

H. Logistics
- Time Frame: 11 to 20+ hours
  The time frames presented here are intended only as a guide. Each teacher has his or her own style, and every class has its own personality and needs. The amount of time invested in each part of the lesson is dependent upon choices the teacher makes about which activities to present, what is assigned for homework or done in class, richness of the discussion, etc.
- Materials to gather (other than copies of handouts): Please see each activity for intended use.
  - Journals (1/person)
  - Dictionaries and Thesauri (1/group of 4 to 5 students)
  - A copy of *The Wizard of Oz* movie
  - Balloons and a stopwatch (note: check for latex allergies when using balloons, non-latex balloons are available; check out http://mrballoon.com/)

relevant standards

What is Collaborative Leadership?

McRel Standards and Benchmarks*:

LIFE SKILLS: WORKING WITH OTHERS

Standard 1: Contributes to the overall effort of a group.

1. Knows the behaviors and skills that contribute to team effectiveness.
2. Works cooperatively within a group to complete tasks, achieve goals, and solve problems.
3. Challenges practices in a group that are not working and proposes measures to enhance team effectiveness.
4. Demonstrates respect for others' rights, feelings, and points of view in a group.
5. Identifies and uses the individual strengths and interests of others to accomplish team goals.
6. Identifies causes of conflict in a group and works cooperatively with others to deal with conflict though negotiation, compromise, and consensus.
7. Helps the group establish goals, taking personal responsibility for accomplishing such goals.
8. Evaluates the overall progress of a group toward a goal.
9. Contributes to the development of a supportive climate in groups.
10. Actively listens to the ideas of others and asks clarifying questions.
11. Takes the initiative in interacting with others.
12. Uses appropriate strategies when making requests of other people.

Standard 2: Uses conflict-resolution techniques.

1. Communicates ideas in a manner that does not irritate others.
8. Identifies individual vs. group or organizational interests in conflicts (e.g., works to build consensus within a group while maintaining minority viewpoints).

Standard 3: Works well with diverse individuals and in diverse situations.

1. Works well with those of the opposite gender, of differing abilities, and from different age groups.
2. Works well with those from different ethnic groups, of different religious orientations, and of cultures different from their own.

Standard 4: Displays effective interpersonal communication skills.

1. Demonstrates appropriate behaviors for relating well with others (e.g., empathy, caring, respect, helping, friendliness, politeness).
2. Exhibits positive character traits toward others, including honesty, fairness, dependability, and integrity.
3. Knows strategies to effectively communicate in a variety of settings (e.g., selects appropriate strategy for audience and situation).
4. Provides feedback in a constructive manner, and recognizes the importance of seeking and receiving constructive feedback in a nondefensive manner.

---

10. Uses emotions appropriately in personal dialogues.
13. Acknowledges the strengths and achievements of others.

Standard 5: Demonstrates leadership skills.

1. Understands one's own role as a leader or follower in various situations.
2. Knows the qualities of good leaders and followers.
3. Knows a variety of leadership strategies, and knows which strategies to implement in specific situations.
4. Demonstrates and applies leadership skills and qualities (e.g., plans wins and celebrates accomplishments; recognizes the contributions of others; passes on authority when appropriate).

LANGUAGE ARTS: WRITING

Standard 4: Gathers and uses information for research purposes.

5. Synthesizes information from multiple research studies to draw conclusions that go beyond those found in any of the individual studies.

LANGUAGE ARTS: LISTENING AND SPEAKING

Standard 8: Uses listening and speaking strategies for different purposes.

5. Makes formal presentations to the class (e.g., includes definitions for clarity; supports main ideas using anecdotes, examples, statistics, analogies, and other evidence; uses visual aids or technology, such as transparencies, slides, electronic media; cites information sources).
8. Responds to questions and feedback about own presentations (e.g., clarifies and defends ideas, expands on a topic, uses logical arguments, modifies organization, evaluates effectiveness, sets goals for future presentations).

LANGUAGE ARTS: VIEWING

Standard 9: Uses viewing skills and strategies to understand and interpret visual media.

1. Uses a range of strategies to interpret visual media (e.g., draws conclusions, makes generalizations, synthesizes materials viewed, refers to images or information in visual media to support point of view, deconstructs media to determine the main idea).
2. Uses a variety of criteria (e.g., clarity, accuracy, effectiveness, bias, relevance of facts) to evaluate informational media (e.g., web sites, documentaries, news programs).
9. Understands how literary forms can be represented in visual narratives (e.g., allegory, parable, analogy, satire, narrative style, characterization, irony).

CIVICS

Standard 9: Understands the importance of Americans sharing and supporting certain values, beliefs, and principles of American constitutional democracy.

3. Understands the significance of fundamental values and principles for the individual and society.

PHYSICAL EDUCATION

Standard 5: Understands the social and personal responsibility associated with participation in physical activity.

1. Uses leadership and follower roles, when appropriate, in accomplishing group goals in physical activities.
4. Includes persons of diverse backgrounds and abilities in physical activity.

LIFE SKILLS: THINKING AND REASONING

Standard 3: Effectively uses mental processes that are based on identifying similarities and differences.

4. Identifies the qualitative and quantitative traits (other than frequency and obvious importance) that can be used to order and classify items.

Standard 6: Applies decision-making techniques.

5. Evaluates major factors that influence personal decisions.

LIFE SKILLS: SELF-REGULATION

Standard 1: Sets and manages goals.

1. Sets explicit long-term goals and shorter range subgoals.
2. Creates an action plan to achieve long-term goals that includes strategic, practical steps and that accounts for the resources needed to achieve these goals.
3. Identifies and monitors resources necessary to achieve a goal.
4. Identifies and ranks relevant options in terms of accomplishing a goal.
5. Prepares and follows a schedule for carrying out options, including contingency plans in the event that the original goal changes or is not met.
6. Maintains an awareness of proximity of a goal by establishing personal milestones.
9. Sets routine goals for improving daily life.
11. Displays a sense of personal direction and purpose.

Standard 4: Demonstrates perseverance.

1. Demonstrates perseverance relative to personal goals.

LIFE SKILLS: LIFE WORK

Standard 7: Displays reliability and a basic work ethic.

1. Understands the concept of reliability (e.g., completing tasks on time; maintaining regular attendance; carrying out assigned tasks; being punctual).
2. Understands ethical character traits as they relate to the workplace (e.g., honesty, integrity, compassion, justice).
4. Knows appropriate behavior for the workplace (e.g., congeniality, collaboration, adaptability, self-control, cooperation, respect for diversity).
7. Gives and receives feedback in a positive manner and requests clarification when needed.

Standard 2: Understands various meanings of social group, general implications of group membership, and different ways that groups function.

1. Understands that while a group may act, hold beliefs, and/or present itself as a cohesive whole, individual members may hold widely varying beliefs, so the behavior of a group may not be predictable from an understanding of each of its members.
2. Understands that social organizations may serve business, political, or social purposes beyond those for which they officially exist, including unstated ones such as excluding certain categories of people from activities.
4. Understands that groups have patterns for preserving and transmitting culture even as they adapt to environmental and/or social change.
5. Understands that social groups may have patterns of behavior, values, beliefs, and attitudes that can help or hinder cross-cultural understanding.

Standard 3: Understands that interactions among learning, inheritance, and physical development affect human behavior.

3. Understands that expectations, moods, and prior experiences of human beings can affect how they interpret new perceptions or ideas.
4. Understands that people might ignore evidence that challenges their beliefs and more readily accept evidence that supports them.

Lesson 3.A Overview: Continue to Intentionally Build and Maintain a Safe Working Environment

Focus:
Create a safe environment for collaboration
Encourage participation
Facilitate brainstorming
Foster a climate of continuous learning
Choose the appropriate facilitation technique

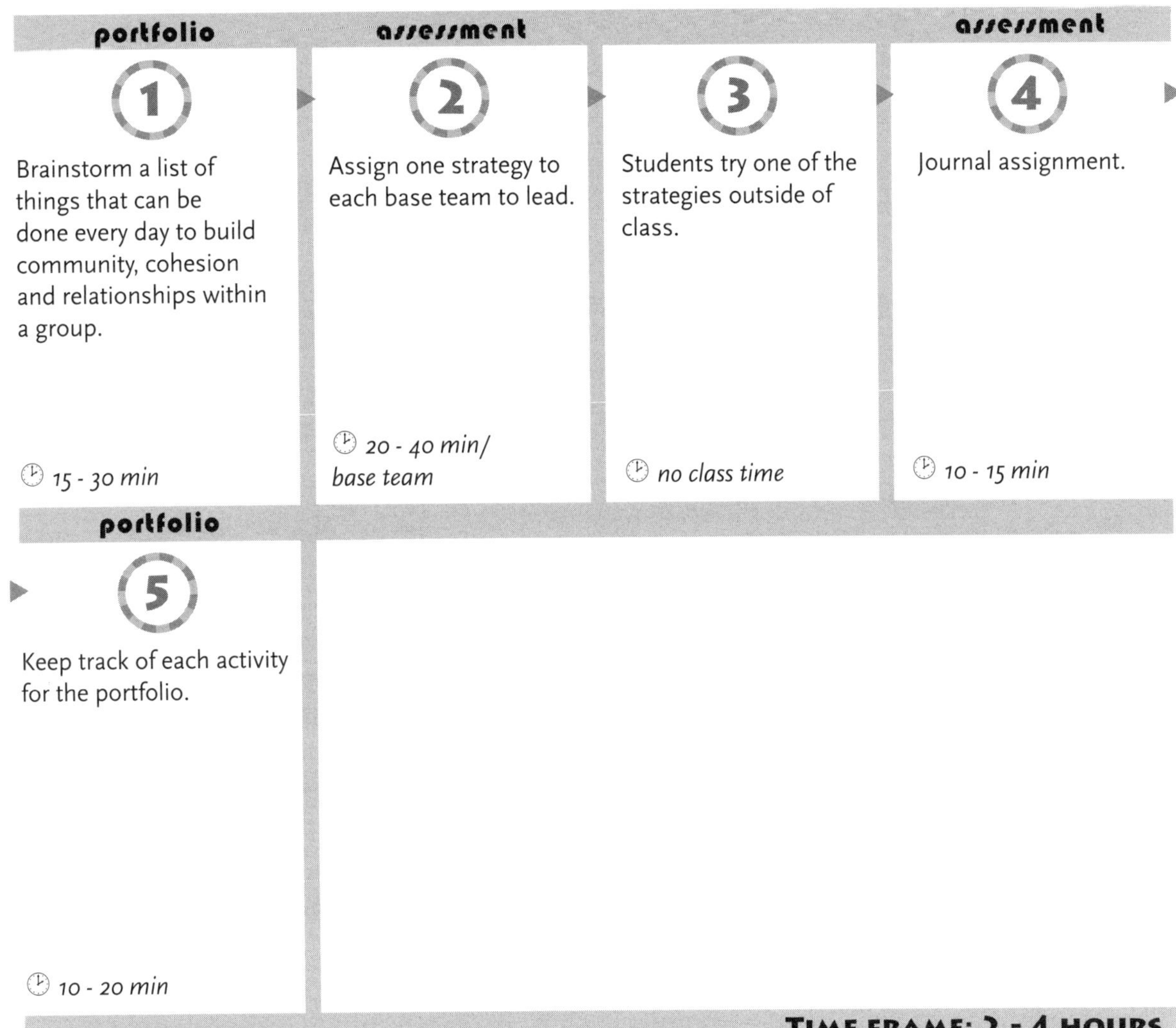

LESSON 3.A SEQUENCE: CONTINUE TO INTENTIONALLY BUILD AND MAINTAIN A SAFE WORKING ENVIRONMENT

Brainstorm a list of things that can be done every day to build community, cohesion, and relationships within a group.

Identify those that can be done quickly and easily. Along with specific and sequenced activities, there are many ways to build cohesion and nurture relationships within a group. Discuss the reason groups come together in the first place:

- Natural groups: family units
- Formal groups: student councils, boards...
- Social groups: parties, gatherings, and just plain hanging out
- Work groups: committees and teams in place of employment
- Volunteer groups: people coming together for a common task
- Unintentional groups: being in a classroom together, or going to camp and being put into a cabin with other people

People end up in groups for all sorts of reasons, either by choice or design. All of these situations provide opportunities for collaborative leadership. If collaborative leadership takes place in the context of relationships, then it is important to think about how to go about creating strong relationships with the people in any group, whatever the reason for the group's existence.

It is not always practical to facilitate a rip-roaring game of Moonball or get out the toys for a Group Juggle. These activities, though, are simply tools to help people come together. What are some other ways to accomplish the same goal? At a social gathering, for example, it may mean that, instead of simply sticking with the people one already knows, you make an effort to mingle with other groups and introduce people to each other. In a work group that will be together over time, you may suggest doing a short icebreaker at the beginning of every meeting to help people get into the meeting and feel more comfortable with sharing. For a short-term meeting group, having everyone introduce him/herself or check in can be enough to help people feel more connected before diving into the tasks at hand.

Brainstorm a list of things that could be done in one or more group situations.

3.A sequence 2

Assign one strategy to each base team to lead.

Choose the ideas that are most appropriate for the class to try. Randomly assign one to each base team to lead over the next week. Every day have one base team lead their strategy with the class. They can be creative about how to do this. For example, if one team gets the strategy to mix at parties, they can stage a mini-party and have people practice their mingling skills.

3.A sequence 3

Students try one of the strategies outside of class.

Have students choose one of the strategies from the original brainstormed list to try with a group outside of class.

3.A sequence 4

Journal assignment.

What strategy for strengthening community/relationships did you try, with whom, and how did it go?

3.A sequence 5

Keep track of each activity for the portfolio.

Have students write down the strategies and keep them in their portfolios.

Lesson 3.B Overview: Assess Personal Collaborative Leadership Values

Focus: Practice reflection
Model collaborative leadership core values of courage, compassion, continuous learning, and service

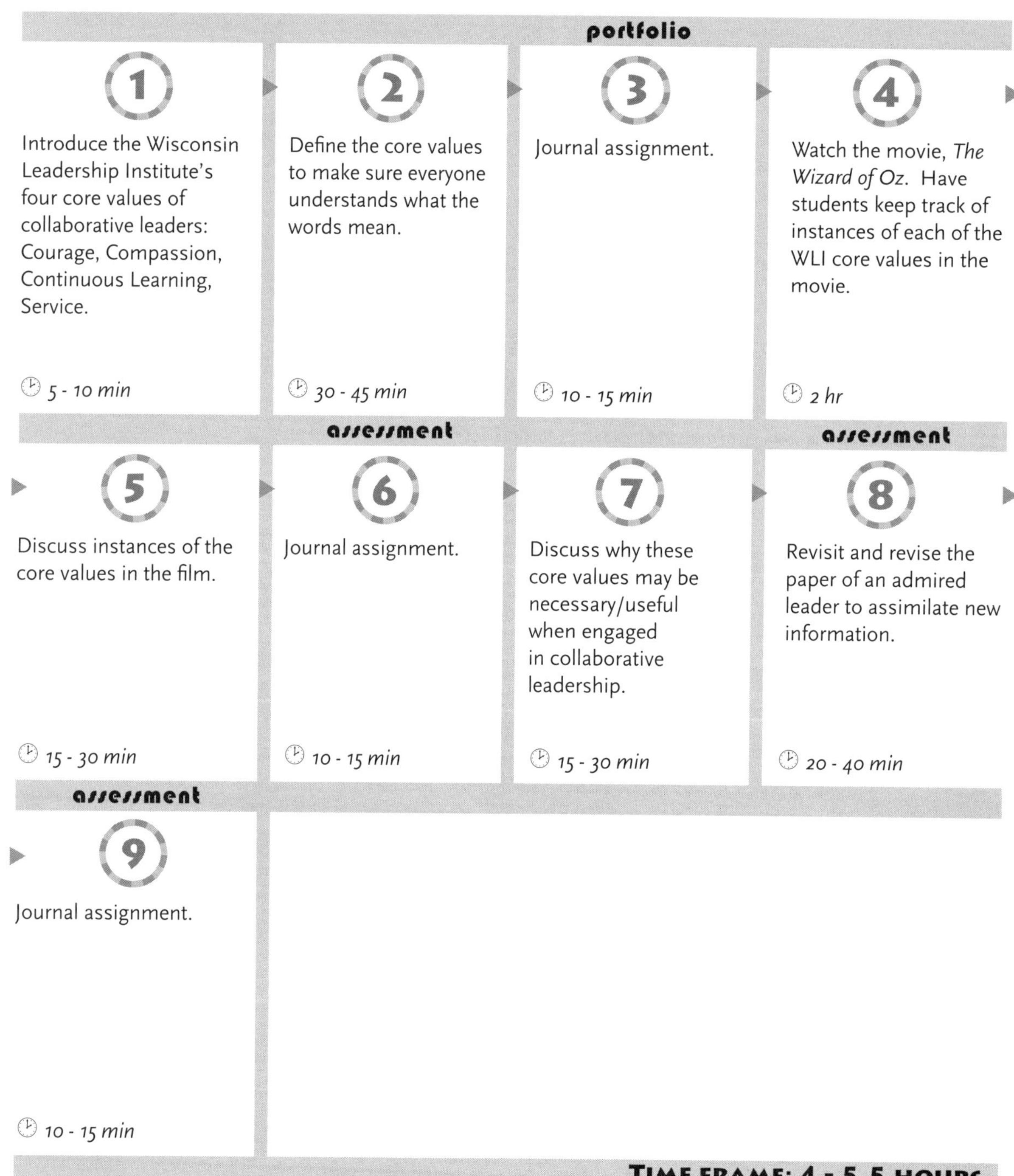

Lesson 3.B Sequence: Continue to Intentionally Build and Maintain a Safe Working Environment

3.B sequence 1

Introduce the Wisconsin Leadership Institute's four core values of collaborative leaders: Courage, Compassion, Continuous Learning, Service.

Introduce each of these concepts as values of collaborative leadership. The following steps of the lesson give students an opportunity to define these, so simply foreshadow that they are important foundations for collaborative leadership.

Make sure students understand that a value is something that people see as having worth and is useful (or "valuable") toward making something happen. In other words, these core values are not things, as much as they are ideas that people see as important.

Define the core values: Courage, Compassion, Continuous Learning, and Service. Make sure everyone understands what the words mean.

activity

Defining WLI Core Values

Focus: Defining the WLI core values of collaborative leadership
Materials: A dictionary and thesaurus for each group (hard copy or computer), writing materials
Time: 30 - 45 minutes
Sequence: Problem Solving
Sources: Original

Suggested Procedure

1. Ask students to get into groups of four.
2. As a large group, have them decide on four things that can be mimed. For example: throwing a Frisbee, swinging a golf club, shooting a basketball, and kicking a soccer ball.
3. In their small groups of four, have them count to 3, and then everyone does one of the actions. The goal is for everyone in each group to do a different action.
4. Keep trying until they do it. (If one or two groups continue to try without getting everyone to be something different, then ask them to make a plan and try again).
5. After all the small groups get everyone to be a different action, then ask all the people doing a given action to group. All the Frisbee throwers group together, the golfers group together, etc. In this way you will have four groups.
6. Assign each group one of the four WLI core values. Give them resources to help define the words they have (dictionary, thesaurus).
7. They have two tasks:
   a) Visually show the class what the words mean. For example, the service people may show helping others, or the continuous learning people may show people taking a knitting class. The actual product is really a side-product of the discussion each group goes through to decide what to do; it is their discussion that is important.
   b) Give a written definition of the words and explain how they fit in with collaboration and leadership.

Sample Processing Questions

- How did you collaborate to arrive at your definition and presentation?
- Did you use any of the core values during this process?
- How might these core values be put into practice in real life?

Facilitation Notes

After the presentations, make sure to point out that these words are abstract concepts as they relate to collaborative leadership. Service, for example can be as much about fixing a car as it is about having a belief that working for the betterment of others and society is important.

3.B sequence 3

Journal assignment.

Write or glue the definitions of courage, compassion, continuous learning, and service in their journals.

Watch the movie, *The Wizard of Oz*. Have students keep track of where they see instances of each of the WLI core values in the movie.

If your school does not already have it, check out a copy of *The Wizard of Oz* from the library or rent it from a video store. This film depicts the WLI four core values very nicely: Courage/Lion; Compassion/Tin Man; Continuous Learning/Scarecrow; Service/Dorothy.

Discuss instances of the core values in the film.

The best places to look for the ways that the various characters represent the core values of courage, compassion, and continuous learning are in two episodes. The first is where the Cowardly Lion, the Tin Man, and the Scarecrow collaborate to rescue Dorothy from the Wicked Witch. The second is the scene where all three receive rewards from the Wizard symbolizing the courage, compassion, and learning they didn't know they had.

As you watch rescue scenes, notice that when it comes time to act together to save Dorothy, the Cowardly Lion achieves courage, the Tin Man's natural caring concern compels him to do what is necessary, and the Scarecrow spontaneously begins cooking up strategies and giving orders for action. As you watch the scene where the Wizard of Oz presents symbolic rewards to Dorothy's three companions, notice how the Wizard's speeches suggest that each of the characters already possessed the qualities they were seeking, and they simply had to recognize those qualities in themselves.

Dorothy herself is the character who most clearly represents the value of service; she spends all of her energy and attention serving the interests of those around her. There are many places in the film where Dorothy demonstrates courage, compassion, and learning as well—all in the pursuit of service. Notice how Dorothy also learns at the end of her journey that she already has the power to achieve the object of her deepest desire, which is to return home. And since the whole story of her adventure takes place in her own dream, it is really the story of Dorothy recognizing her own courage, compassion, capacity for learning, and devotion to service.

3.B sequence 6

Journal assignment.

Choose a character in *The Wizard of Oz*. How did this character show one or more of the WLI core values of collaborative leaders?

3.B sequence

Discuss the Wisconsin Leadership Institute's four core values of collaborative leaders.

Discuss how and why the Wisconsin Leadership Institute's four core values may be necessary/useful when engaged in collaborative leadership.

Courage: Leaders always have to stand up and speak up for the good of the group, which can sometimes be scary. Never be afraid to take sensible risks; ultimately the riskiest strategy is to play it safe. If you fall down, get up, just as you did when you were learning to walk. Never be afraid to speak truth to people in power. If it won't kill you, it will probably make you stronger. If you wonder whether a risk is worth taking, ask yourself the following questions:

- What is the worst thing that can happen?
- What is the best thing that can happen?
- What is most likely to happen?
- How much of the outcome is under your control?
- How much effort are you willing to spend to make it work?
- How will the outcomes affect those around you?

Compassion: Caring about other people is often the main reason why good leaders get started as leaders. We are all linked by empathy and caring concern. Nobody can survive from birth in isolation. Every one of us started out small, helpless, and dependent on the care of others. Leaders are followers are leaders. We need each other in many ways. Your heart is as smart as your brain.

Continuous learning: Nothing succeeds like success, but nothing teaches like mistakes. The most effective leaders avoid the word "failure." They speak instead of "glitches," "bollixes," and "snafus," all of which are merely opportunities for learning. Leadership is a problem-solving activity; if we want to solve problems, all we really need to do is to learn what it is we need to learn and then learn it.

Service: The best leaders are servants. If you serve others in need, they will serve you in your times of need. All leaders serve some higher purpose. What higher purpose will you serve?

3.B sequence

Revisit and revise the paper of an admired leader to assimilate new information (from Lesson 2.B, p. 171).

Add a section about how this person does or does not embody the core values of a collaborative leader.

3.B sequence

Journal assignment.

Write about ways you have put these core values into action in your life. Write about ways you would like to put these values into action in your life.

Lesson 3.c Overview: Use Collaborative Leadership Qualities and Values

Focus:

Encourage participation
Make sure that people are heard
Facilitate problem solving
Model collaborative leadership values of courage, compassion, continuous learning, and service
Practice reflection

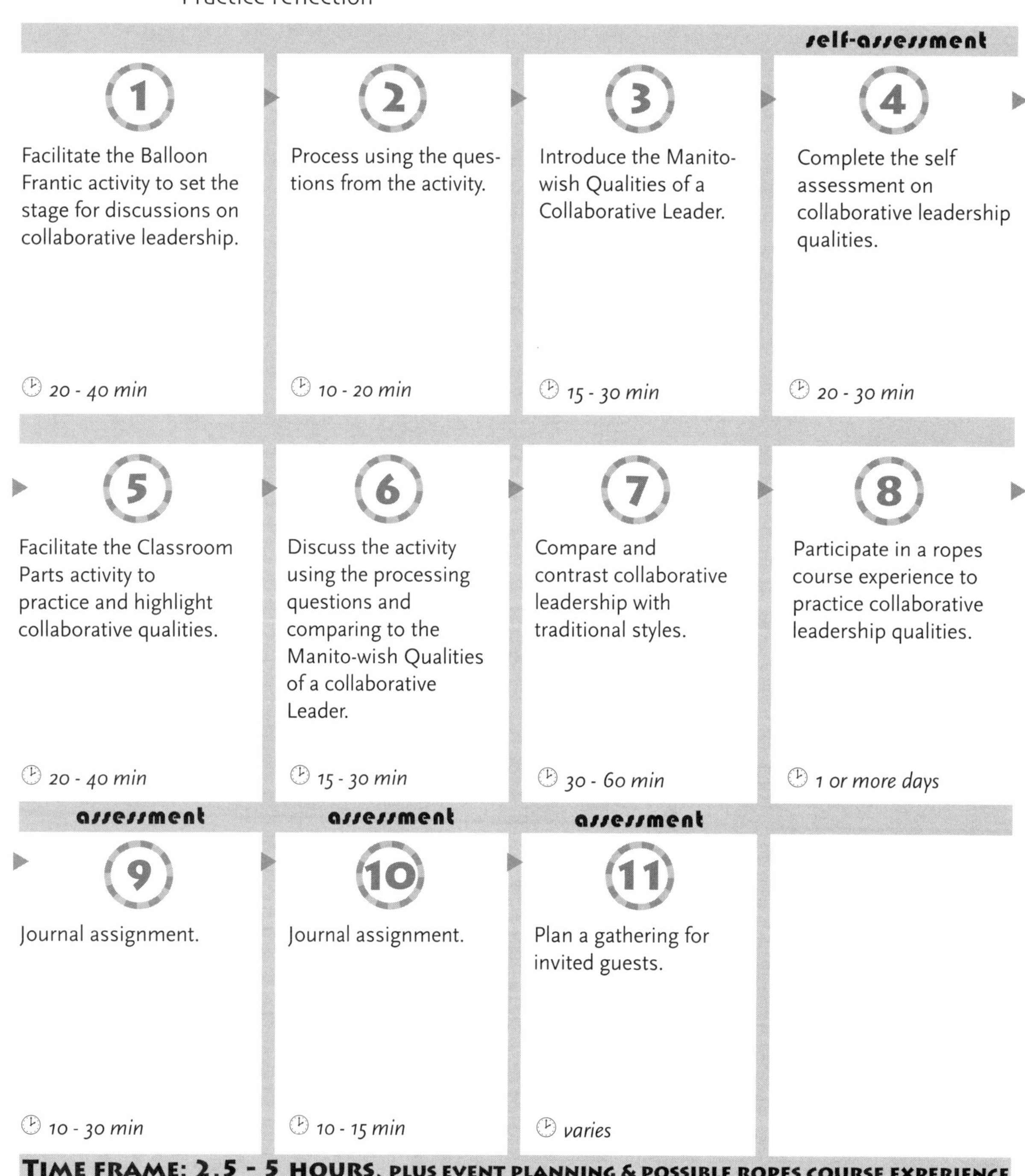

Time Frame: 2.5 - 5 Hours, plus event planning & possible ropes course experience

LESSON 3.C SEQUENCE: CONTINUE TO INTENTIONALLY BUILD AND MAINTAIN A SAFE WORKING ENVIRONMENT

3.C sequence 1

Facilitate the Balloon Frantic activity to set the stage for discussions on collaborative leadership.

The purpose for doing this activity is to set the stage for the conversations about collaborative leadership that will ensue, and make connections between the values and the qualities of collaborative leadership.

activity

BALLOON FRANTIC

Focus: Cooperation, problem solving, goal setting, conflict resolution, leadership, organization, decision making

Materials: Balloons and a stopwatch (note: check for latex allergies when using balloons, non-latex balloons are available; check out http://mrballoon.com/)

Time: 20 - 40 minutes

Sequence: Problem Solving

Sources: *Journey Toward the Caring Classroom* by Frank, "Equally Frantic" in *Diversity in Action* by Chappelle & Bigman, "Balloon Bash" in *Games for Teacher* by Cavert & Frank, and *Silver Bullets* by Rohnke.

Suggested Procedure

1. You will need 2 or 3 balloons per person (large balloons are best), and a stopwatch.
2. Ask participants to blow up their balloons as large as they want and tie them off.
3. Everyone keeps one balloon, and the rest are piled in a space near the group.
4. On a signal, everyone begins bouncing his/her balloon in the air. The balloons must be bounced, not held.
5. Every 5 seconds, add another balloon.
6. The object is to see how long the group can keep balloons bouncing before receiving 6 penalties.
7. A balloon hitting the floor is a penalty. Once on the floor, if the balloon is not back in play within 5 seconds, that is another penalty.
8. Every time a penalty is assessed, the facilitator shouts out which number it is. So, when the first balloon touches the floor, the facilitator shouts, "One!" If another one touches the floor, or that same balloon sits for five seconds, the facilitator shouts, "Two!" When the facilitator gets to "six," the time is stopped.
9. After some discussion time, the group attempts to better its record.

Sample Processing Questions

- How did your strategies (models) change with each attempt?
- As you became more experienced with this task, what did you do to share the responsibilities?
- Describe some of the feelings you had when it got frantic?
- What caused it to be frantic? As a group, how did you get that under control?
- What are some responsibilities you have in your life?
- What are some strategies you use to juggle all of the responsibilities in your life?
- Do you ever feel overwhelmed with the responsibilities?
- When working in groups, how can you share responsibilities?

- How do you know when a group is collaborating successfully?
- How did you work together to make this successful?

Facilitation Notes

This activity can get frantic during the first few attempts, and a group rarely makes it very far. Once a group agrees to organize, they then can make it much longer.

It helps to have two people facilitating this activity—one to add balloons every five seconds, the other to watch for penalties. The number of attempts can either be open ended, with the group deciding when they have achieved their best time (i.e., setting a goal for themselves), or they can have a finite number of attempts. Generally, a group needs at least five attempts in order to fine-tune their strategies.

Balloon Frantic is a wonderful metaphor for the frantic pace of many people's lives. Participants can label their balloons with some of the responsibilities they have in life. As the group organizes, they see that other people can help them with some of their responsibilities, instead of having to juggle everything themselves.

Process using the questions from the activity.

Using the processing questions from the activity, discuss how the activity went, and how people showed any or all of the four core values of a collaborative leader.

Introduce the Manito-wish Qualities of a Collaborative Leader.

Camp Manito-wish YMCA has been in existence for almost 90 years. Over the years, Manito-wish staff noticed that may of the former campers grew to become noted leaders in their communities, families, and businesses. It also became clear that the type of leadership these alumni were practicing was different than the norm. John Stanley, Director of Camp Manito-wish during the 1990's, researched this phenomenon. He determined that the Manito-wish experience prompted young people to develop leadership skills. This experience involved groups of campers spending days together preparing and carrying out wilderness trips, which ranged in time from 2 days to almost 2 months. Because these young people did everything in, with, and for the group, the leadership that was developing was collaborative in nature.

With this insight, John researched the concept of Collaborative Leadership. One of his many resources was *No More Teams!* by Michael Schrage. From this and other books, John developed the 7 Qualities of a Manito-wish Leader. Once articulated, these 7 Qualities are now intentionally focused upon so that young people can mindfully develop these skills as they interact and work together through their wilderness trips. They then generalize these skills to other parts of their lives.

The idea behind introducing these qualities is to simply introduce them. Deeper interpretation and application of these qualities occurs through activities and reflection. Every person will integrate them into their own wealth of experiences, which means that they may interpret them differently. Throughout the course ask questions so that students can continue to evolve in their understanding of these qualities and how they can use them in a multitude of situations.

Seven Qualities of a Manito-wish Collaborative Leader

- **Builds a shared vision with those they lead.**
  A clear vision motivates others to commit to an idea and follow it through. It encourages others to collaborate toward the realization of that vision. What is it you want to accomplish together? What is your vision and what are the visions of others? How might they be combined?

- **Builds models—tries it... changes it... tries it again.**
  A trial-and-error approach, and the making of mistakes, is considered part of the continuous learning experience. If people are serious about what they do, it is necessary to build lots of models and be prepared for success, failure, or something in-between. It is risk-taking in action, and it takes courage.

- **Shares a common space with others.**
  If group members are occasionally passing each other in their daily routine, dropping each other notes, e-mail, or simply talking on the phone, it all adds up to simple communication, not collaboration. Communication is good, but it is not enough. Face-to-face contact is essential if effective collaboration is to take place. At some point, the group must meet in one place in order to throw ideas around and make important decisions.

  Another aspect of sharing a common space is the idea of allowing everyone an opportunity to speak. Some people are naturally quieter than others. They must make an effort to speak up, and the more vocal must allow pauses in the conversation for the quieter folks to have an opening.

- **Lets others amplify their abilities.**
  Competition can get in the way of collaboration, so it is necessary for group members to leave their egos at the door. If everyone is included and they are able to draw on the strengths of everyone in the group, a synergy can develop that transcends the individuals. In other words 1+1=3.

- **Remembers ... followership and leadership go hand-in-hand.**
  Sometimes it is an act of leadership to take a step back and allow others to take the lead. Viewing "followers" as adversaries to be controlled is counter-productive. Seeing others as allies with unique gifts empowers everyone to contribute to the shared vision.

- **Doesn't collaborate to turn out the lights.**
  During a crisis, collaborating is dangerous. Autocratic leadership is what is called for in a crisis situation. Likewise, collaborating to complete a simple task is a waste of energy and resources. Sometimes, one person taking the initiative is enough. The term "analysis paralysis" is often used when a group realizes it is engaged in unnecessary collaboration.

- **Celebrates successful collaborations.**
  This is as important as celebrating individual accomplishments. Celebrating is so important and a terrific investment in the next challenge which demands that a group collaborate to achieve the vision. Theater groups have a cast party on the final night of the show. This is a good example of celebrating collaborations.

Complete the self assessment on collaborative leadership qualities (on the following page).[2]

For reference, the statements on the self assessment are meant to correspond with the following Manito-wish Qualities of a Collaborative Leader.

- Builds a shared vision with those they lead: B, H
- Builds models—tries it... changes it... tries it again: I, M
- Shares a common space with others: A, C, G, J, L
- Lets others amplify their abilities: E, G, J
- Remembers ... followership and leadership go hand-in-hand: C, D, G
- Doesn't collaborate to turn out the lights: D, I, K
- Celebrates successful collaborations: F

SELF-ASSESSMENT: QUALITIES OF COLLABORATIVE LEADERSHIP

The following statements are based on the qualities of collaborative leadership. Circle the number that most closely reflects your attitude or behavior when working in (or with) a group of people.

NEVER	SOMETIMES	OFTEN	ALWAYS
1	2	3	4

A. When something comes up that calls for communication between another and myself, I prefer that we meet and talk about it.

1 2 3 4

B. In discussion, I refer to our shared vision whenever possible, in order to bring the group together, keep us on task, or remind us of our goal.

1 2 3 4

C. When we have a group task to work on, I try to avoid competing with others and/or placing obstacles in the way of the group.

1 2 3 4

D. When someone seems to be doing most of the work in a group task, I try to contribute more toward its completion.

1 2 3 4

E. In a disagreement with another member of the group, I listen respectfully and try to appreciate why he or she has that opinion.

1 2 3 4

F. I congratulate my group members when they have succeeded in something or have done a good job.

1 2 3 4

G. I am willing to hold back my opinion when I realize that I've been talking a lot.

1 2 3 4

H. I like to bring up the goals of our meeting when the discussion gets bogged down in details that are not relevant to those goals.

1 2 3 4

I. I can change my mind (or the direction it's going in) when getting the task done seems to call for it.

1 2 3 4

J. When I notice that someone is especially quiet in a group discussion, I ask for his or her perspective or opinion.

1 2 3 4

K. When it is not necessary to collaborate on a task, I either do it myself, or ask someone else to do it.

1 2 3 4

L. When a decision or problem arises that affects the group, I bring it to the group to discuss.

1 2 3 4

M. When a task or a project is complete, I take time to talk with the group about how things went.

1 2 3 4

Facilitate the Classroom Parts activity to practice and highlight collaborative qualities.
Do this activity to explore the Manito-wish Qualities of a Collaborative Leader. After the activity, focus in on the qualities. Where did the students notice a quality was evident. Which ones were not displayed during this activity?

activity

Classroom Parts

Focus: Creating a shared vision, working together on a task, trust
Materials: A deflated balloon or small beach ball for each group of four or five. Check for latex allergies. Use non-latex balloons, if possible (see http://mrballoon.com/)
Time: 20 - 40 minutes
Sequence: Problem Solving
Sources: Chris Cavert & Jackie Gerstein in *Games for Teachers* by Cavert & Frank.

Suggested Procedure

1. Clear the furniture from the middle of the room
2. Divide the class into groups of 4 to 5.
3. Give each group a small beach ball, or give each group a deflated balloon if you are concerned about latex allergies.
4. Explain that their task is to blow up the balloon, cross to the other side of the room, and break the balloon (or deflate the beach ball), in that order. This is done as a unit, using the following:
    - four legs (five legs for groups of 5)
    - three hands
    - two eyes
    - one mouth
5. They are to choose who plays each of the roles (e.g., who are the four legs, the three hands, etc.)
6. Once they choose who plays which part, then only the person playing the "mouth" can talk.
7. As a unit, they must figure out how to blow up the balloon, tie it off, cross the room, and pop the balloon.
8. While doing this, they must be in constant and continual contact with one another in their unit.
9. If they break contact in any way, they go back to the beginning, re-choose roles and try again.
10. Ask the groups to quietly decide how much planning time they would like (up to 10 minutes), and write it on a piece of paper. Whatever they decide is how much time they get. If they choose to not use all of the time, they may save up to 3 minutes for planning in case their attempt fails and they need to return to the beginning. (In this way, each group may have a different amount of planning time).

Sample Processing Questions

- On a scale of 1 to 10 (1 being easiest, 10 being most difficult), how challenging do you think this activity was?
- What made it challenging/not challenging for you? What was the most challenging part?
- Which leadership qualities did you use, or notice others using, to accomplish this task?
- Which qualities did you not see used during the activity?
- Did you use integrity while doing this activity (e.g., follow the stated guidelines), or did you fudge on the rules a bit?

Facilitation Notes
This can be a very challenging activity, and you must be alert for safety issues. Caution participants to take it slow and make sure they are really working together before lifting and while carrying anyone. If it is too chaotic and there are safety issues, take time-out to talk about your concerns. If it continues, stop the activity altogether.

The challenging nature of this activity does mean people have to collaborate in order to have any success with it. No one can do this alone. One thing to look out for is whether or not people are adhering to the guidelines. You may need to give some reminders.

Note that this activity is not a competition. One strategy is to total the time it takes for everyone to complete the task and then see if they can reduce the time. This will give the entire class a goal. Better yet, have the students decide how to make it less of a competition, and more of a collaboration.

Discuss the activity using the processing questions and comparing to the Manito-wish Qualities of a Collaborative Leader.

Compare and contrast collaborative leadership with traditional styles.

Have a discussion, comparing and contrasting collaborative leadership with traditional styles (autocratic, democratic, laissez faire) and situational leadership® (see pp. 190-191 for a review of traditional and situational leadership). One of the main differences between traditional styles of leadership and collaborative leadership is the positioning of the individual. In a traditional view of leadership, the leader stands alone in a position of authority. In a collaborative approach, the individual is a leader among leaders. The group is the anchor from which leadership takes place depending upon the strengths of individuals, the tasks and problems that need to be tackled, and the needs of the group. One or more people may have positions of authority within the group, but it is simply one of the many roles this person(s) takes on.

Situational leadership takes on new meaning when practicing collaborative leadership. Rather than one individual judging how to act depending on the skills and needs of everyone else, the group of people openly discuss needs and skills to ascertain what needs to be done in a given situation. In this way, leadership percolates within the stew that is the group. Each person steps up to take on leadership at any given point depending on how their skills match the needs at that moment. This amplifying of each other's abilities is difficult without being on the same page by creating the shared vision, understanding that followership is as important as leadership, and intentionally building the relationships for all of the collaborative leadership qualities to thrive.

See pp. 4-5 for a chart comparing collaborative leadership and traditional leadership behaviors.

Another way to compare the two is through the following definitions put forward by Jack Christ:

> **Leadership** is a reciprocal process of encouraging and supporting... people in the pursuit of goals shared by members of a group, organization, or community.

Collaborative leadership is a reciprocal process of encouraging and supporting... relationships within which people can pursue a variety of shared goals over extended periods of time.

These definitions point out the subtle difference between the two perspectives on leadership, which can also be summed up in the Manito-wish Qualities of a Collaborative Leader. Spend some time deconstructing these definitions and how they fit with definitions that were created be each student.

Participate in a ropes course experience to practice collaborative leadership qualities.
Consider having a ropes course experience to practice collaborative leadership using the qualities and risk taking as themes.

If a local ropes course provider is unknown, there are three organizations that can help you find a course in your area:

Association for Experiential Education (AEE): 866-522-8337, www.aee.org.
Association for Challenge Course Technology (ACCT): 847-325-5860, www.acctinfo.org

The answers to the following questions can help determine if a ropes course provider can meet your safety and programming needs.

- **Who built your course?**
  These days, the Association for Challenge Course Technology (ACCT) is recognized as a leader in determining standards for ropes/challenge course construction. A course that is built by ACCT Professional Vendor Members means that a certain level of quality and safety standards has been met. Sometimes organizations build their own courses, and then have it inspected by an ACCT Professional Vendor Member to verify that the standards have been met.
- **When was the last time your course was inspected?**
  ACCT recommends a yearly inspection by an ACCT Professional Vendor Member, in addition to periodic inspections by the people who run the course. If the course has not been inspected by an outside source in a while, ask why. If the organization has not had their course inspected, or does not feel it is important, consider taking your group elsewhere. An outside inspection is a small price to pay to ensure the physical safety of your class.
- **What kind of training do you require for your staff?**
  Typically, staff engage in an adventure skills workshop from a reputable vendor that includes facilitation techniques, technical/safety skills, and program philosophy. Some staff have this training when they are hired, while others go through the training after being hired. In addition, initial orientation for course protocols and periodic refreshers are given to staff so they can maintain their skills. More training and staff development is a bonus. There should be at least one person with first aid and CPR certification working on the course while groups are there. If you are doing high elements, there should be at least one person on the course who is trained in rescue techniques.
- **Is your program accredited?**
  Much like AAA for hotels and dining and American Camping Association (ACA) for camps, the Association for Experiential Education (AEE) offers an accreditation program for ropes and challenge courses. Using industry standards, they have teams visit courses to look over entire

programs. When a course is subsequently accredited, it is a sign that certain minimum standards are met. Accreditation is voluntary and is not required. There are many ropes/challenge course programs out there that far exceed minimum industry standards yet are not accredited.

- **What is your plan in case of injury?**
  This should be a no-brainer. If they cannot rattle off what to do in case of an emergency, be wary.
- **Describe a typical day at your course.**
  Look for signs that they pay attention to the sequencing of activities, and that they are prepared to alter their plan depending upon the needs of the group.
- **What is your organization's philosophy on Challenge by Choice?**
  Ropes/challenge courses regularly put participants in vulnerable situations. Participants need to be allowed to choose their own level of challenge. This does not mean that facilitators take a hands-off approach, because there are times when students need an emotional nudge or encouragement to push themselves. It does mean that staff have thought about this in advance and understand that individuals react differently to anxiety. A one-size-fits-all approach is problematic.
- **Do you have any ideas about how to help us transfer this experience back at school?**
  Although they are not responsible for connecting learning from this short experience back to the classroom, it is nice to know that they are thinking about it. Some organizations have strategies to help with this.
- **How do you deal with conflicts between group members?**
  Conflict is a natural part of the group process. Chances are that your students will experience some level of conflict during the day at a ropes/challenge course. Have they thought about it? You know your students much better than a facilitator who just met them, which can cause some awkward moments during a conflict situation. Are you ready to defer to them, and are they ready to enlist your help, if necessary?

Although there are many more questions that could be listed here, these questions will give you the ability to "feel-out" the ropes/challenge course provider. If you can find the right provider to meet your needs, your students will have a more powerful experience.[3]

3.C sequence

Journal assignment.

Define each collaborative leadership quality in your own words.

3.C sequence

Journal assignment.

Revisit leadership definition and modify as necessary.

Plan a gathering for invited guests.

As a class, plan a gathering (lunch, teach-in, team building hour, workshop, appreciation time...). Brainstorm ideas and prioritize by using a voting process. To vote on numerous ideas, you can give each person more than one vote. Count up the ideas and divide by 3. That is how many votes each person gets. For example, if there are 15 ideas, each person gets five votes. An individual can then put all their votes on one item, or spread them out among the items as they wish. If two or more items have an almost equal number of votes, then have a run-off with everyone voting between the items.

Next remind students of the collaborative leadership qualities. Give the class a certain amount of time to plan the event, and let it roll. Observe how students collaborate ... or not. Make notes about which qualities are being used ... or not. Periodically check in with students and guide them with questions about their process. See if they would like to continue as they are or make changes.

After the event, remember to celebrate the collaboration and do a final debrief of the process using the qualities and core values.

Lesson 3.D Overview: Practice Effective Use of Collaborative Leadership Core Values and Qualities

Focus:
Encourage participation
Practice reflection
Develop capacity to act in the face of risk and uncertainty
Foster creativity

1	2 (assessment)	3 (assessment)	4
Review the unit skills and themes. Look for any questions or gaps.	Create an action plan for how to lead collaboratively in one's own life.	Students write a letter to themselves about their class experience and goals.	Plan and carry out a small class service project to benefit the school or local community.
15 - 30 min	*30 - 60 min*	*15 - 30 min*	*varies*

5 (assessment)	6 (assessment)	7 (assessment)	8 (assessment)
Debrief and evaluate the project.	Students chart portfolio completion using the chart provided.	Students and teacher evaluate portfolio.	Evaluate the course.
15 - 30 min	*20 - 40 min*	*varies*	*15 - 30 min*

Time Frame: 2 - 4 Hours, plus planning & carry out of service project

Lesson 3.D Sequence: Continue to Intentionally Build and Maintain a Safe Working Environment

1 Review the course skills and themes. Check to see if there are any questions or gaps.

Unit 1: What is Collaboration? Skills/Themes	Unit 2: What is Leadership? Skills/Themes	Unit 3: What is Collaborative Leadership? Skills/Themes
Developing Group Cohesion • Group building activities • Shackleton's Voyage • Into Thin Air • Qualities of productive and non-productive teams • Social contract	**Creating Community** • Types of activities • Sequencing activities • Facilitating activities	**Intentional Community Building** • Strategies to build relationships
Diversity • Multiple intelligences • Being an ally • Group identities • Honoring diversity	**Personal Definition of Leadership** • Leader role model • Qualities of a leader • Compose a working definition of leadership	**Collaborative Leadership Values** • WLI core values: courage, compassion, continuous learning, service
Collaboration • Definitions of Cooperation and Collaboration • Similarities and differences between cooperation and collaboration	**Are Leaders Born or Made?** • Aristocracy and Democracy • Debate: are leaders born or made?	**Qualities of a Collaborative Leader** • Manito-wish 7 Qualities: build a shared vision, build models, share a common space, let others amplify abilities, followership and leadership go hand-in-hand, doesn't collaborate to turn out the lights, celebrate successful collaborations • Collaborative leadership qualities self assessment • Ropes/challenge course experience (if applicable) • Gathering for invited guests
Skills/Tools for Collaboration • Communication: active listening, non-verbal cues • Decision-making: types of decisions, brainstorming, consensus, STEPS process • Conflict Resolution: recognizing conflict, conflict escalation and anger, win-win solutions, STEPS process • Goal Setting: group goals, nominal group process, personal goals, SMART goals	**Views of Leadership** • Autocratic, Democratic, Laissez Faire • Situational leadership® • Collage on an admired leader	**Practicing Collaborative Leadership** • Action plan for your own life • Write a letter to yourself • Plan and carry out a service project
	Risk Taking and Vision • Risk taking • Vision and organization • Personal vision and mission	

Action Plan for Collaborative Leadership

1. Choose a group in which to practice collaborative leadership values, skills, and qualities (family, club, peer group, student council, etc.).

2. Think about all of the values, skills, and qualities of a collaborative leader we have explored in this course (check out your portfolio). What are five skills, qualities, and/or values that jump out at you as being important?

3. Choose one to develop over the next month or two. Write a SMART GOAL for how you would like to go about developing it with your chosen group.

4. What are some possible barriers that may get in the way of you achieving your goal?

5. What can you do to deal with these barriers (prevent them from happening, or deal with them once they happen)?

6. How will you acknowledge the success of achieving your goal (e.g., call a friend, treat yourself to something, etc.)?

3.D sequence 3

Students write letters to themselves about their class experience and goals.
Have students write letters to themselves reflecting on their class experience and the SMART goal they just identified for themselves. Have them place the letters in a self-addressed stamped envelope for you to send out on an agreed-upon date. It is important that you have a plan to do this. If you are not available to send the letters on the agreed-upon date, make sure you have a back-up plan.

3.D sequence 4

Plan and carry out a small class service project to benefit the school or local community.
Ask a student to co-facilitate the nominal group process with you in order to brainstorm and identify a targeted, short-term service project that will benefit the school or local community. The scope of the project will be determined by the energy level of the students, commitment level of everyone, and amount of time left for this course.

Make sure that the chosen project does not infringe on anyone's personal beliefs and values. Coming to consensus is important for commitment from the entire class.

A wonderful resource for this is *The Kid's Guide to Service Projects* by Barbara A. Lewis.

Once a project has been identified, create committees for each part of the project. Create a shared vision and set group goals, individual goals, and a timeline for what needs to be done. Decide how the class will evaluate the project (and their roles in the evaluation) and how they will celebrate when the project is over. Carry out the project.

3.D sequence 5

Debrief and evaluate the project (don't forget to celebrate!).
Ask students to develop an evaluation as part of their planning process.

3.D sequence 6

Students chart portfolio completion using the chart provided.
Using the following chart, students determine what assignments are in their portfolio for this unit and reflect on them by adding comments. The purpose of this exercise is to help students remember what they have done and reflect on their learning. This is a list of the unit assignments and assessments. There can certainly be more items in the portfolio than are listed here, especially if a student wishes to use it to keep track of his/her work.

PORTFOLIO COMPLETION CHART

Item (Lesson)	Comments
List of things that can be done to build community, cohesion, and relationships. (3.A)	
Journal: Try community, cohesion relationship strategy on your own. What strategy did you try, with whom, and how did it go? (3.A)	
Journal: Wisconsin Leadership Institute definitions of their four core values for collaborative leadership: Courage, compassion, continuous learning, service. (3.B)	
Journal: Choose a character in the film, *The Wizard of Oz*. How did this character show one or more of the WLI Core Values of collaborative leaders? (3.B)	
Journal: Write about ways you have put these core values into action in your life. Write about ways you would like to put these values into action in your life. (3.B)	
Self assessment of collaborative leadership qualities. (3.C)	
Journal: Define each collaborative leadership quality in your own words. (3.C)	
Journal: Revisit leadership definition and modify as necessary. (3.C)	

Portfolio Completion Chart

Item (Lesson)	Comments
Action plan for how to lead collaboratively in one's own life. (3.D)	
A letter to yourself about this class experience and your goals (it will be sent to you).	

3.D sequence 7

Students and teacher evaluate portfolio.

Evaluation will depend on the purpose of the class and the culture in the school. One way to evaluate portfolios is to have each student convene a group of 3 peers. Student then choose a sampling of their work to share with this review committee. The committee is to look over the work and meet with the student to ask questions and provide feedback about the quality of the work. Before doing this, you may wish to create some ground rules about feedback and how it will be offered. Using your established social contract is an option, as well.

The same sampling of work can then be evaluated by the student him/herself and the teacher. If necessary, grades can be suggested by each of the parties, and a grade can be assessed for the portfolio.

Evaluate the course.

Decide on criteria for the course evaluation with the whole class. Ask students to evaluate the courses using their journals, portfolios, and grid of skills & themes. This can be done anonymously in a written report and/or publicly in a discussion format using small groups or the entire class.

---

[1] Adapted from Frank, L. & Stanley (1997) *Manito-wish Leaders' Manual: Teacher Edition*, Boulder Junction, WI: The Manito-wish YMCA (no longer in print).

[2] Adapted from *Self Assessment in Collaborative Leadership* by Ruth A. Gudinas, FULL CIRCLE, N9136 Big Lake Road, Gresham, WI 54128-8955, 715/787-4427, with help from Laurie Frank.

[3] From Frank, L. (2004). *Journey Toward The Caring Classroom* (p. 262). Used with permission.

"A SENSE OF HUMOR IS PART OF THE ART OF LEADERSHIP, OF GETTING ALONG WITH PEOPLE, AND GETTING THINGS DONE."

—DWIGHT D. EISENHOWER

Activities

Journal Questions

Lesson 1.A

p. 33: Describe your experience of getting to know the people in this group. What were your feelings when you came into this class on the first day? What are they now?

Lesson 1.B

p. 68: Choose one or more: How has your uniqueness been helpful or hurtful to you in school and in your life? How have you been an ally, and/or wish to be an ally with others? Who do you look down upon and how would your life be different if you decided to become their ally? What have you been picked on for and what would you like your allies to say or do about this?

Lesson 1.C

p. 81: Using their base team definitions of collaboration have each student rewrite the definition to create a personal definition of collaboration.

Lesson 1.D

p. 104: What made the Pathfinder Consent activity easy and/or frustrating for you? How did you like having to check in with everyone (and be checked in with) before anyone could take a step?

p. 111: What did you learn about how decisions are made in your life? Do you want to change how you make decisions? If so, how? If not, how is it working for you?

p. 118: What is the difference between needing something and wanting something? How can one's wants cause conflict? How can one's needs cause conflict?

p. 123: Which of the de-escalators works best (or might work best) for you? How will you know to use the de-escalators the next time one of your "triggers" is pushed? Record your reaction in your journal.

p. 125: Bring back the scenario from the beginning of the unit and ask students to brainstorm win-win solutions:

> *The night before you asked your brother if you could use the bathroom first in the morning because you had to leave early for a school trip. He agreed. When you woke up, he was in the bathroom, and taking his time, too. You pound on the door and say ". . ."*

Lesson 1.E

p. 139: Look at your group and personal goals for this project. How did you do? Did you achieve your goals? Explain.

Lesson 2.B

p. 175: Using the list of traits brainstormed by the class and the life skills that support leadership, write about the traits you feel are strong for you, and the ones you would like to develop more.

p. 175: Write a personal definition of leadership. As more information and insights are gained-about leadership, they will have an opportunity to modify this definition.

Lesson 2.C

p. 184: Revisit your working definition of leadership and modify it as necessary.

Lesson 2.D

p. 192: Revisit your working definition of leadership and modify it as necessary.

Lesson 2.E

p. 195: How might risk-taking relate to you when in a leadership role?

Lesson 3.A

p. 217: What strategy for strengthening community/relationships did you try, with whom, and how did it go?

Lesson 3.B

p. 220 Write or glue the definitions of courage, compassion, continuous learning, and service in their journals.

p. 221 Choose a character in *The Wizard of Oz*. How did this character show one or more of the WLI core values of collaborative leaders?

p. 221 Write about ways you have put these core values into action in your life. Write about ways you would like to put these values into action in your life.

Lesson 3.C

p. 231 Define each collaborative leadership quality in your own words.

p. 231 Revisit leadership definition and modify as necessary.

Bibliography

Albin, J. (2004). *Changing the message: A handbook for experiential prevention.* Oklahoma City, OK: Wood 'N' Barnes Publishing.

Carlin, C. (n.d.). *Leadership dynamics one training manual.* Williams Bay, WI: Lake Geneva-Genoa City Union High School District.

Bennis, Warren G. (1994). *On becoming a leader.* Reading, MA: Perseus Publishing.

Blanck, W.(2001). *The 108 skills of natural born leaders.* New York, NY: AMACOM, American Management Association.

Brendtro, L. K. & Larson, S. J. (2006). *The resilience revolution: Discovering strengths in challenging kids.* Bloomington, IN: Solution Tree formerly National Education Service.

Brown Lehr, J. & Martin, C. (1994). *Schools without fear: Group activities for building community.* Minneapolis, MN: Educational Media Corporation.

Burke, K. (1999). *How to assess authentic learning.* Arlington Heights, IL: SkyLight Professional Development.

Cain, J. & Joliff, B. (1998). *Teamwork and teamplay: A guide to cooperative, challenge adventure activities.* Dubuque, IA: Kendall/Hunt Publishing Co.

Cavert, C. (1999). *Affordable portables: Working-book of initiative activities & problem solving elements – revised and expanded.* Oklahoma City, OK: Wood 'N' Barnes Publishing.

Cavert, C. (1999). *Games (& other stuff) for group, book 1, rev. and exp. ed.).* Oklahoma City, OK: Wood 'N' Barnes Publishing.

Cavert, C. & Frank, L. (1999). *Games (& other stuff) for teachers: Classroom activities that promote pro-social learning.* Oklahoma City, OK: Wood 'N' Barnes Publishing.

Chappelle, S. & Bigman, L. (1998). *Diversity in action: Using adventure activities to explore issues of diversity with middle school and high school age youth.* Hamilton, MA: Project Adventure, Inc.

Conzemius, A., & O'Neill, J. (2001). *Building shared responsibility for student learning.* Alexandria, VA: Association for Supervision & Curriculum Development.

Conzemius, A. & O'Neill, J. (2002). *The handbook for SMART school teams.* Bloomington, IN: Solution Tree.

DePree, M. (1993). *Leadership jazz: The art of conducting business through leadership, followership, teamwork, voice, touch.* New York: Dell Publishing.

Dodds, T., & Prosser-Dodds, L. (2003). *Games for change: Group activities with creative spiritual concepts on the side.* Oklahoma City, OK: Wood 'N' Barnes Publishing.

Frank, L. (2004). *Journey toward the caring classroom: Creating community in the classroom and beyond.* Oklahoma City, OK: Wood 'N' Barnes Publishing.

Frank, L. & Panico, A. (2007). *Adventure education for the classroom community.* Bloomington, IN: Solution Tree.

Frank, L. S., & Stanley, J. (1997). *Manito-wish leaders manual: Teacher ed.* Waukesha, WI: The Manito-wish YMCA.

Gardner, H. (1999). *Intelligence reframed: Multiple intelligences for the 21st Century.* New York: BasicBooks.

Gibbs, J. (2000). *TRIBES: A new way of learning and being together.* Sausalito, CA: CenterSource Systems.

Glasser, W. (1988). *Choice theory in the classroom.* New York: HarperCollins Publishers, Inc.

Gregson, B. (1982). *The incredible indoor games book.* Belmont, CA: David S. Lake, Publishers.

Henton, M. (1996). *Adventure in the classroom: Using adventure to create a community of life-long learners.* Hamilton, MA: Project Adventure, Inc.

Hersey & Blanchard. *Center for leadership studies: Home of situational leadership®*. http://www.situational.com/

Krakauer, J. (1997) *Into thin air: A Personal Account of the Mount Everest Disaster*. New York, NY: Random House.

Krakauer (September, 1996). Into thin air. *Outside Magazine*. http://outside.away.com/outside/destinations/199609/199609_into_thin_air_1.html

Kreidler, W. J. & Furlong, L. (1996). *Adventures in peacemaking: A conflict resolution activity guide*. Cambridge, MA: Educators for Social Responsibility & Project Adventure, Inc.

Kreidler, W. J. (1984). *Creative conflict resolution*. Glenview, IL: Scott, Foresman and Company.

Lansing, A. (1999). *Endurance: Shackleton's Incredible Voyage, 2nd ed*. New York: Carroll & Graf Publishers.

Lantieri, L., Lieber, C. M., & Roderick, T. (1998). *Conflict resolution in the high school*. Cambridge, MA: Educators for Social Responsibility.

Levin, M. (2007). Unpublished Manuscript. www.marilynlevin.com.

Lewin, K., Lippitt, R., & White, R. K. (1939). Patterns of aggressive behavior in experimentally created social climates. *Journal of Social Psychology, 10*, 271-279.

Lewis, B. A. (1995). *The kid's guide to service projects*. Minneapolis, MN: Free Spirit Publishing.

Lewis, B. A. (1998). *What do you stand for?: A kid's guide to building character*. Minneapolis, MN: Free Spirit Publishing.

Lewis, B. A. (1998). *The kid's guide to social action*. Minneapolis, MN: Free Spirit Publishing.

MacGregor, M. G. (2007). *Everyday leadership in all teens*. Minneapolis, MN: Free Spirit Publishing.

Maxwell, J. (1999). *21 indispensable qualities of a leader*. Nashville, TN: Thomas Nelson, Inc.

Mayer, B. (2000). *The dynamics of conflict resolution: A practitioner's guide*. San Francisco: Jossey-Bass, Inc.

Morrell, M., Capparell, S., & Shackleton, A. (2002). *Shackleton's way: Leadership lessons from the great Arctic explorer*. New York: Penguin Books.

Nova. (2002, March 26). *Shackleton's Voyage of Endurance*. Retrieved March 19, 2004 from http://www.pbs.org/wgbh/nova/transcripts/2906_shacklet.html

Olson, K. & Pearson, S (2000). *Character begins at home: Family tools for teaching character and values*. Kent,WA: Susan Kovalik and Associates, Inc. Available from Books For Educators Inc.

Roehlkepartain, J. L. (1997). *Building assets together: 135 group activities for helping youth succeed*. Minneapolis, MN: The Search Institute.

Rohnke, K. (2004). *Funn 'n games*. Dubuque, IA: Kendall Hunt Publishing.

Rohnke, K. & Butler, S. (1995) Quicksilver. Dubuque, IA: Kendal Hunt Publishing.

Rohnke, K. (1989). *Cowstails and cobras II*. Dubuque, IA: Kendall/Hunt Publishing Company.

Rohnke, K. (1984). *Silver bullets: A guide to initiative problems, adventure, games and trust activities*. Dubuque, IA: Kendall/Hunt Publishing Company.

Seuss, Dr. (1961). *The Sneetches and other stories*. New York: Random House.

INDEX

AUTHORS

Laurie Frank is a former public school teacher who has worked in the adventure and experiential education arenas for 25 years. She began her career as a special education teacher in emotional disabilities, working with students of all ages. Her path diverged upon the discovery of adventure education and experiential methodologies. The need to develop community within the school setting was apparent, and the adventure philosophy seemed the perfect vehicle to achieve that goal.

Laurie is the owner/director of GOAL Consulting, working with school districts, camps and non-profit organizations around the country to create environments where students, faculty, staff, and families are invited into the educational process. She also helps schools develop experiential education curriculum for children and young adults.

Ms. Frank was a leader in designing the nationally recognized Stress/Challenge adventure program for Madison (Wisconsin) Metropolitan School District, and wrote their curriculum, "Adventure in the Classroom," in 1988 and she wrote Journey Toward the Caring Classroom (2004), an experiential approach to creating community in schools. She also collaborated on two books: one with Chris Cavert, titled Games (& Other Stuff) for Teachers (1999), and the other with Ambrose Panico titled, Adventure Education for the Classroom Community (2000, Revised 2007).

Laurie has been a Certified Trainer with Project Adventure, and is a recipient of the Michael Stratton Practitioner of the Year award from the Association for Experiential Education. She currently serves on the boards of the Association for Experiential Education, Wisconsin Youth Company, Wil-Mar Neighborhood Center, and the Hancock Center for Movement Arts and Therapies. In 2008 she became part of a collective of owners with Wood 'N' Barnes Publishing dedicated to providing quality books by respected authors on experiential approaches to learning.

Carol Ruth Carlin is a career public school teacher who has always employed active learning techniques in her classroom. She began her career as an English/Social Studies teacher, progressed through speech and the graphic arts and into leadership development. Her journey included participation as an adult leadership trainer for both the state and national teacher's associations' peer leadership training programs. It was that experience that lead Carol to understand the power of collaborative leadership training to empower people to discover the value of their own strengths and contributions. That experience, along with a cooperative educational team, allowed her to develop a highly successful high school residential leadership training program.

Carol is currently semi-retired, teaching at both the community college and teacher graduate studies level. She continues to volunteer in the community. The opportunity to work with the development of this collaborative leadership curriculum has been a delight. The genuine collaborative style exhibited by Laurie Frank, the key author and project coordinator, has been both validating and inspirational.

Dr. Jack Christ earned a Ph.D. from the University of Pennsylvania and joined the faculty of Ripon College in 1970. In 1980, he founded the Leadership Studies Program at Ripon College, the world's first undergraduate liberal-arts academic program offering a minor in Leadership Studies. In 1994, he wrote the report for a Wisconsin Governor's Commission which led to the creation of the Wisconsin Leadership Institute (WLI) and its primary initiative, the Collaborative Leadership Network (CLN). He has served as WLI Executive Director since 1996. He helped create the Ethical Leadership Program at Ripon College in 2004 and currently serves as its Research Director. He is President and CEO of two corporations: Video Age Productions, which produces video programs for the educational marketplace and for public television; and Pluribus, which is devoted to demonstrating the practical value of ethical leadership, global collaboration, and life-long learning. Jack and his wife Bev have two adult children and several pets.